I0488224

NIST Special Publication 800-130

A Framework for Designing Cryptographic Key Management Systems

Elaine Barker
Computer Security Division
Information Technology Laboratory

Miles Smid
Orion Security Solutions
Silver, Spring, MD

Dennis Branstad
NIST Consultant
Austin, TX

Santosh Chokhani
Cygnacom
McLean, VA

August 2013

U.S. Department of Commerce
Penny Pritzker, Secretary

National Institute of Standards and Technology
Patrick D. Gallagher, Under Secretary of Commerce for Standards and Technology and Director

Authority

This publication has been developed by NIST to further its statutory responsibilities under the Federal Information Security Management Act (FISMA), Public Law (P.L.) 107-347. NIST is responsible for developing information security standards and guidelines, including minimum requirements for Federal information systems, but such standards and guidelines shall not apply to national security systems without the express approval of appropriate Federal officials exercising policy authority over such systems. This guideline is consistent with the requirements of the Office of Management and Budget (OMB) Circular A-130, Section 8b(3), *Securing Agency Information Systems*, as analyzed in Circular A-130, Appendix IV: *Analysis of Key Sections*. Supplemental information is provided in Circular A-130, Appendix III, *Security of Federal Automated Information Resources*.

Nothing in this publication should be taken to contradict the standards and guidelines made mandatory and binding on Federal agencies by the Secretary of Commerce under statutory authority. Nor should these guidelines be interpreted as altering or superseding the existing authorities of the Secretary of Commerce, Director of the OMB, or any other Federal official. This publication may be used by nongovernmental organizations on a voluntary basis and is not subject to copyright in the United States. Attribution would, however, be appreciated by NIST.

Comments on this publication may be submitted to:

National Institute of Standards and Technology
Attn: Computer Security Division, Information Technology Laboratory
100 Bureau Drive (Mail Stop 8930) Gaithersburg, MD 20899-8930
Email: ckmsdesignframework@nist.gov

Reports on Computer Systems Technology

The Information Technology Laboratory (ITL) at the National Institute of Standards and Technology (NIST) promotes the U.S. economy and public welfare by providing technical leadership for the Nation's measurement and standards infrastructure. ITL develops tests, test methods, reference data, proof of concept implementations, and technical analyses to advance the development and productive use of information technology. ITL's responsibilities include the development of management, administrative, technical, and physical standards and guidelines for the cost-effective security and privacy of other than national security-related information in Federal information systems. The Special Publication 800-series reports on ITL's research, guidelines, and outreach efforts in information system security, and its collaborative activities with industry, government, and academic organizations.

Abstract

This Framework for Designing Cryptographic Key Management Systems (CKMS) contains topics that should be considered by a CKMS designer when developing a CKMS design specification. For each topic, there are one or more documentation requirements that need to be addressed by the design specification. Thus, any CKMS that addresses each of these requirements would have a design specification that is compliant with this Framework.

Keywords

access control; confidentiality; cryptographic key management system; cryptographic keys; framework; integrity; key management policies; key metadata; source authentication.

Acknowledgements

The National Institute of Standards and Technology (NIST) gratefully acknowledges and appreciates contributions by all those who participated in the creation, review, and publication of this Framework. NIST also thanks the many contributions by the public and private sectors whose thoughtful and constructive comments improved the quality and usefulness of this publication. Many useful suggestions that were made during the workshops held on CKMS at NIST in 2009, 2010, and 2012 have been incorporated in this document.

Contents

Figures

Tables

1. Introduction

This Framework for Designing Cryptographic Key Management Systems (CKMS[1]) is a description of the topics to be considered and the documentation requirements (henceforth referred to as requirements) to be addressed when designing a CKMS. The CKMS designer satisfies the requirements by selecting the policies, procedures, components (hardware, software, and firmware), and devices (groups of components) to be incorporated into the CKMS, and then specifying how these items are employed to meet the requirements of this Framework.

A CKMS consists of policies, procedures, components and devices that are used to protect, manage and distribute cryptographic keys and certain specific information, called (associated) metadata herein. A CKMS includes all devices or sub-systems that can access an unencrypted key or its metadata. Encrypted keys and their cryptographically protected (bound) metadata can be handled by computers and transmitted through communications systems and stored in media that are not considered to be part of a CKMS.

This CKMS Framework provides design documentation requirements for any CKMS. In other words, it describes what needs to be documented in the CKMS design. The goal of the Framework is to guide the CKMS designer in creating a complete uniform specification of the CKMS that can be used to build, procure, and evaluate the desired CKMS.

This Framework offers the following advantages:
a) It helps define the CKMS design task by requiring the specification of significant CKMS capabilities,
b) It encourages CKMS designers to consider the factors needed in a comprehensive CKMS,
c) It encourages CKMS designers to consider factors and mechanisms that, if properly addressed, can provide security to the system,
d) It can be used when logically comparing different compliant CKMS systems and their capabilities,
e) It aids in performing a security assessment by requiring the specification of implemented and supported CKMS capabilities, and
f) It forms the basis for a U.S. Federal CKMS Profile.

NIST Standards and Special Publications are referenced in this Framework as examples only. This Framework is intended to be general enough to encompass any reasonably complete and well-designed CKMS.

---

[1] CKMS can be either singular or plural in this document and should be read as such.

This Framework is not intended to be a CKMS design. That task is left to the CKMS designers. Rather, the Framework provides specification requirements using lists of options that the designers may choose to incorporate in their design.

This Framework specifies documentation requirements, not security requirements. It does not mandate particular security features. The requirements of this Framework are placed on the CKMS design documentation. The Framework aids the designer by providing the essential implementation choices that form the basis of a good CKMS design. The specific choices that ensure a secure CKMS are left to the designer or to other documents, such as security profiles that are based on this Framework.

This Framework does not mandate requirements for the protection of the information belonging to a given public or private sector (e.g., the U.S. Government, the financial industry, or health care services). It is anticipated that sectors will either develop their own profiles, or they will adopt the profiles of other sectors that fulfill their own requirements.

Requirements for conformance to this Framework are indicated by a "**shall**" statement. Recommendations are indicated by the use of "should", but are not requirements for compliance with this Framework. The words "must" or "need(s) to" convey assumptions upon which this Framework is based, but do not constitute a specific requirement on the CKMS design documentation. In this Framework, "**FR:*i.j*"** indicates the jth Framework Requirement in Section *i*.

FR:1.1 A conformant CKMS design **shall** meet all "**shall**" requirements of the Framework.

Since the requirements in this Framework are documentation requirements, it may be adequate to address a requirement by stating that the feature specified in the requirement is not implemented in the CKMS. In many requirements, the words "if, how, where, and under what circumstances" may appear. The "if" indicates a conditional requirement. If the answer to the "if" question is "no" then the designer is expected to address the requirement by indicating why the condition does not apply. If the answer to the "if" question is "yes", then the designer is expected to address the requirement by providing the information levied by the requirement. The "how" response should address how the requirement is met (i.e., how it will be implemented, enforced, and used). The "where" response should address where (logically in the system) the implementing mechanism is located. Finally, the "under what circumstances" response addresses the conditions that must apply before the mechanism is used.

A CKMS design that adequately addresses, specifies, and satisfies all the requirements specified herein can be considered as conforming to, and complying with, this Framework. A conformant CKMS design can be compared to another conformant CKMS design by examining the design specifications meeting each requirement.

1.1 Scope of this Framework

A CKMS will be a part of a larger information system that executes information processing applications. While the CKMS supports these applications by providing cryptographic key management services, the particular applications or particular classes of applications are beyond the scope of this Framework.

Some introductory material is provided to describe the Framework topics and to justify the requirements; however, this Framework assumes that the reader has a working knowledge of the principles of key management or is able to find that information elsewhere (e.g., in [SP 800-57-part1]). Appendix A contains a list of references that are useful in understanding cryptography and cryptographic key management and their application to information security.

1.2 Audience

This Framework is primarily intended for CKMS designers. However, it may also be used by anyone interested in a Cryptographic Key Management System design and related design specifications. It is anticipated that CKMS security analysts, procurement officials, implementers, integrators, operators, and responsible managers would be interested in the CKMS design specifications and products conforming to this Framework.

CKMS designers are expected to use this Framework as a checklist for addressing all the topics covered, for considering all the aspects of a comprehensive CKMS, for selecting those policies, components, and devices to be included in a CKMS, for specifying all the decisions made in the design, and for documenting the decisions with detailed specifications and justifications. The resulting design documentation should be adequate for implementers to create the product, for integrators to incorporate the product in other products or sub-systems, and for procurement officials to understand, evaluate, and compare the product with others having similar characteristics.

1.3 Organization

Section 1 Introduction provides an introduction to the Key Management Framework and the motivation behind it.

Section 2 Framework Basics covers basic concepts of this Framework and provides an overview of the Framework.

Section 3 Goals defines the goals of a robust CKMS.

Section 4 Security Policies discusses the structure, typical contents, and need for information management, information security, CKMS security, and other related security policies.

Section 5 Roles and Responsibilities presents the roles and responsibilities that support a CKMS.

Section 6 Cryptographic Keys and Metadata covers the most critical elements of a CKMS: keys and metadata, by enumerating and defining possible key types; key metadata; and key and metadata management functions, along with access control considerations, security issues and recovery mechanisms.

Section 7 Interoperability and Transitioning considers the need for interoperability and the ability to easily make transitions in CKMS capabilities in order to accommodate future needs.

Section 8 Security Controls describes security controls applicable to a typical CKMS.

Section 9 Testing and System Assurances describes security testing and assurances.

Section 10 Disaster Recovery deals with disaster recovery in general and of a CKMS specifically.

Section 11 Security Assessment discusses the security assessment of a CKMS.

Section 12 Technology Challenges briefly discusses the technical challenges provided by new attacks on cryptographic algorithms, key establishment protocols, CKMS devices, and quantum computing.

Appendix A enumerates and describes useful references.

Appendix B consists of a glossary of terms used in this Framework.

Appendix C provides a list of acronyms used in this Framework.

2. Framework Basics

This section discusses the motivation, intent, properties, and limitations of a Cryptographic Key Management Framework.

2.1 Rationale for Cryptographic Key Management

Today's information systems and the information that they contain are considered to be critical assets that require protection. The information used by government and business is often contained in computer systems consisting of groups of interconnected computers that make use of shared networks, e.g., the Internet. Since the Internet is shared by diverse and often competing organizations and individuals, information systems should protect themselves and the information that they contain from unauthorized disclosure, modification, and use. In addition, denial of service to legitimate users could be considered a significant threat in many service and time-critical application systems and the CKMS used to protect them. Additional security requirements can be derived from the organizational goals for protecting personal privacy, including anonymity, unlinkability, and unobservability of CKMS-supported communications. The information used by these systems requires protection when it is at rest, when it is being processed within a protected facility, and also when it is transported from one location to another.

Cryptography is often used to protect information from unauthorized disclosure, to detect unauthorized modification, and to authenticate the identities of system entities (e.g., individuals, organizations, devices or processes). Cryptography is particularly useful when data transmission or entity authentication occurs over communications networks for which physical means of protection (i.e., physical security techniques) are often cost-prohibitive or even impossible to implement. Thus, cryptography is widely used when business is conducted or when sensitive information is transmitted over the Internet. Cryptography can also provide a layer of protection against insiders and hackers who may have physical or possibly logical access to stored data, but not the authorization to know or modify the data (e.g., maintenance personnel or CKMS users).

Cryptographic techniques use cryptographic keys that are managed and protected throughout their lifecycles by a CKMS. Effectively implemented cryptography can reduce the scope of the information management problem from the need to protect large amounts of information to the need to protect only keys and certain metadata (i.e., information about the key and its authorized uses, such as the algorithm with which the key is to be used, the security service to be provided using the key, etc.).

When designing a CKMS, the cryptographic techniques used to protect the keys managed by the CKMS should offer a level of protection called the security strength that is infeasible for a would-be attacker to bypass or subvert; the security strength of the technique is the base 2 logarithm of the minimum number of operations required to cryptanalyze the algorithm, and is often measured in bits of security. This design principle is comparable to a design principle used in building safes and vaults: the designer builds the vault to a standard that would discourage a rational attacker from attempting entry; the only feasible way to open the safe is to open the safe door by trying possible combinations until the correct combination is selected. Similarly, the only way to decrypt previously encrypted data (without knowledge of the correct key) is to test possible keys until, eventually, the correct key is used to decrypt the ciphertext to obtain the correct plaintext. Just as the protection provided by a safe is dependent on the number of its possible combinations, the strength of a cryptographic algorithm is dependent on the number of possible keys.

Other means of gaining access to the contents of the safe or to the information that has been encrypted may also exist. One can drill through the safe enclosure, and one can attempt to find a shortcut method to cryptanalyze the cryptographic algorithm. Also, one can attempt to steal the correct combination or key. Safe combinations and cryptographic keys both require similar protection. The CKMS should be designed to provide the necessary protection for keys and metadata.

Cryptography can be used to provide three major types of protection to data: confidentiality, integrity, and source authentication.

a) *Confidentiality* protection protects data from unauthorized disclosure. Encryption algorithms are used to transform plaintext data into unintelligible ciphertext, while decryption algorithms are used to transform the ciphertext back to the original plaintext. The transformations are controlled by one or more cryptographic keys so that only the authorized parties who have the keys can successfully perform the transformations.

b) *Integrity* protection provides mechanisms to detect unauthorized data modifications. Cryptographic authentication algorithms typically calculate an authentication code or digital signature, which is a function of the data being protected and a cryptographic key used by the algorithm. It is highly unlikely that without possession of the correct key, an entity could modify the data and compute the correct authentication code or digital signature. Therefore, unauthorized modifications of data can be detected before the modified data is used.

c) *Source authentication* provides assurance that the protected data came from an authorized entity. For example, suppose that a digital signature is calculated on data and is transmitted with the data. The receiver can verify the digital signature and therefore know that the data came from a particular entity. In this Framework, source authentication involves authenticating the identity of the source and then verifying that the authenticated entity is authorized to participate in the function being performed.

These protections can be provided to any data protected by the CKMS, including keys and the associated metadata (See Section 6.2.1, items s) and t)).

Cryptographic algorithms should reside within a cryptographic module (consisting of hardware, software, firmware, or a combination thereof) which physically and logically protects its contents (e.g., the algorithms, cryptographic keys, and metadata) from unauthorized modification and disclosure. A cryptographic module is part of a CKMS and can provide cryptographic protections to keys, metadata, and user data.

FR:2.1 The CKMS design **shall** specify all cryptographic algorithms and supported key sizes for each algorithm used by the system.

FR:2.2 The CKMS design **shall** specify the estimated security strength of each cryptographic technique that is employed to protect keys and their bound metadata.

2.2 Keys, Metadata, Trusted Associations, and Bindings

A key must be associated with metadata that specifies characteristics, constraints, acceptable uses, and parameters applicable to the key. For example, a key may be associated with metadata that specifies the key type, how it was generated, when it was generated, its owner's identifier, the algorithm for which it is intended, and its cryptoperiod. Each unit of metadata is called a metadata element. Like keys, metadata needs to be protected from unauthorized modification and may need to be protected from disclosure; the metadata also needs to have its source adequately authenticated.

A metadata element may be implicitly known and therefore may not be specifically recorded for certain keys within a CKMS. For example, if all keys within a device are AES-128 keys, then a metadata element recording key sizes may not be required. However, in many systems, there is a need to differentiate one key from another using one or more explicitly recorded metadata elements. This CKMS framework focuses on those metadata elements that are explicitly recorded and managed by the CKMS. The term "metadata" is used in this context (i.e., the term "metadata" refers to explicitly recorded and managed metadata elements).

There are many possible metadata elements for a given key. A *trusted association*, between a key and selected metadata elements, is often needed by the CKMS in order to perform key management functions. For example, it is desirable to have a trusted association between a static public key and the owner's identifier. When used in conjunction with an owner registration process, the trusted association provides assurance that the owner that is specified by the identifier is, or was, in possession of the corresponding private key.

Metadata elements may be generated by the same entity that generates the key, or they may be received from a trusted entity. Whenever metadata is received from a trusted entity (whether or not the associated key is sent simultaneously) there must be a trusted association between the metadata and the associated key. The trusted association maintained during the distribution may be enforced by a *cryptographic binding* (*binding*) of the key and metadata (e.g., a digital signature computed on the combination of the key and metadata), or the association may be enforced by a *trusted process* (e.g., a face-to-face handover of metadata from an entity who is known and trusted). A CKMS often provides cryptographic binding and verification functions that are used in the key and metadata distribution and management processes. The receiver obtains assurance that the key and its metadata are properly associated, have come from a particular source, have not been modified, and have been protected from unauthorized disclosure during transit. Upon receipt of the metadata, the association between the key and metadata should be verified. A cryptographic binding is verified by applying the appropriate cryptographic verification function to the key and bound metadata elements. A non-cryptographic trusted association is verified by assessing the trusted process (i.e., the trust in the sending entity and the distribution process). See Figure 1 below.

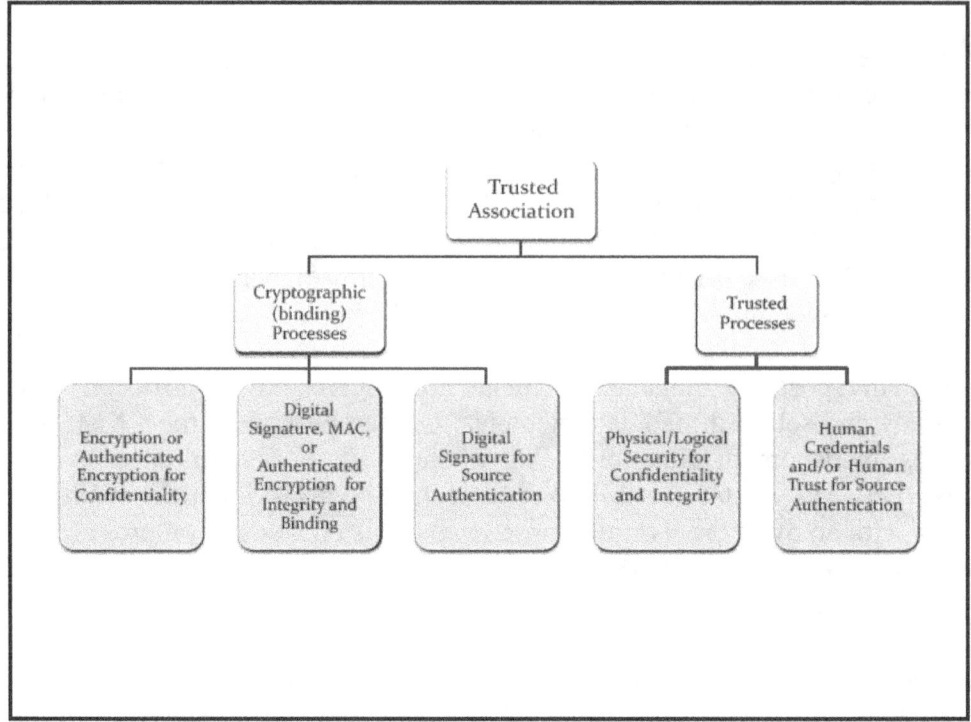

Figure 1: The Trusted Association and Supporting Processes

After being received, the metadata can be combined with other locally generated metadata (if available), and a new trusted association between the key and all available metadata can then be established for the information to be stored.

Metadata stored within a system also needs a trusted association between the key and its metadata. Depending on the storage location and characteristics, the association could be maintained using physical security or cryptographic methods. Physical security methods include storage within a device that is trusted to maintain the association, i.e., the confidentiality (when required) and the integrity of a key and its metadata. As long as the integrity of the trusted association is maintained, one has assurance that the metadata elements belong to the associated key and have not been disclosed to unauthorized entities. However, such physical security methods might not be feasible. A physically secure storage site might be too costly or might not be available. In this case, a cryptographic binding could be required to provide assurance that a key and its metadata are properly associated.

2.3 CKMS Applications

A CKMS can be designed to provide services for a single individual (e.g., in a personal data storage system), an organization (e.g., in a secure VPN for intra-office communications), or a large complex of organizations (e.g., in secure communications for the U.S. Government). A CKMS can be owned or rented.

2.4 Framework Topics and Requirements

This Framework contains a list of Framework Topics (FTs) (corresponding to the section headings) and, for each topic, a set of Framework Requirements (FRs) that need to be satisfied when designing a CKMS (see Figure 2 below). These requirements are placed on the CKMS design.

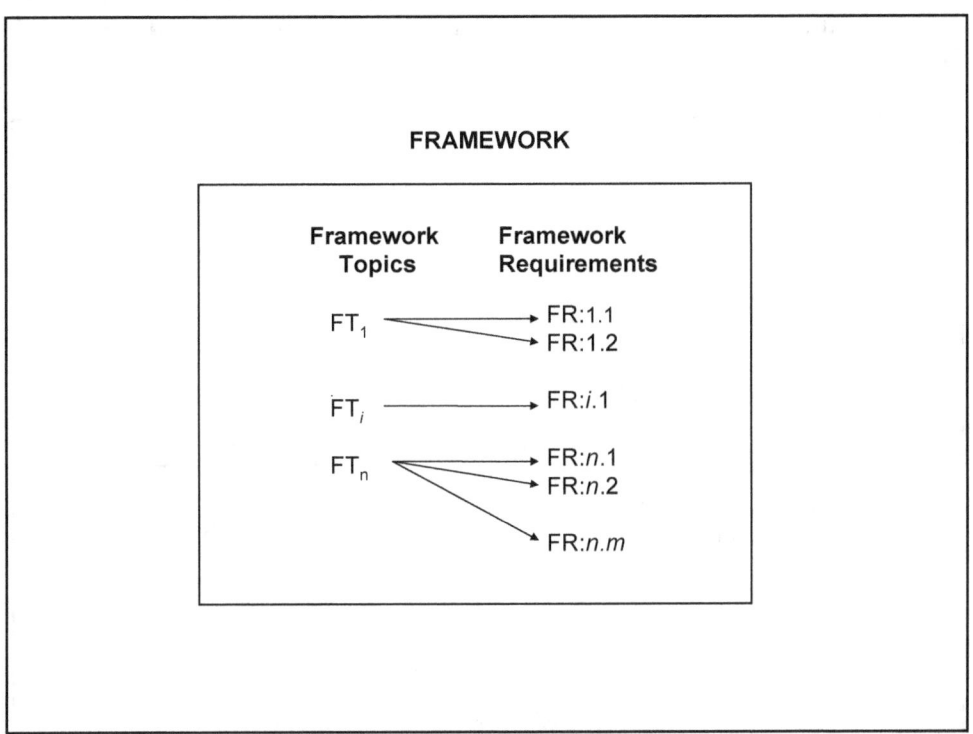

Figure 2: Framework of Topics and Requirements

This Framework does not impose any specific policies, procedures, security requirements, or system design constraints on the CKMS; it simply requires that they be documented in a structured manner so that CKMS designs can be understood and compared.

This Framework is not oriented to a particular CKMS or class of CKMS for a sector (such as the U.S. Federal Government, Aerospace, Health Care, etc.). This Framework is intended to be applicable to all CKMS.

FR:2.3 A compliant CKMS design **shall** describe design selections and provide documentation as required by the requirements of this Framework.

2.5 CKMS Design

The purpose of a CKMS design is to describe how a system can be built to provide cryptographic keys to the entities that will use those keys to protect sensitive data. The high-level description of the CKMS should indicate the uses of each key type, where and how keys are generated, how they are protected in storage at each entity where they reside and during delivery, and the types of entities to whom they are delivered.

Figure 3 illustrates how a CKMS Design can be shown to be compliant with this Framework. For each Framework requirement **FR:*i.j***, the appropriate Framework response, **fr:*i.j***, is provided by the CKMS designer to meet the requirement. The complete set of pairs consisting of requirements and responses {**FR:*i.j***, **fr:*i.j***} form the CKMS Design.

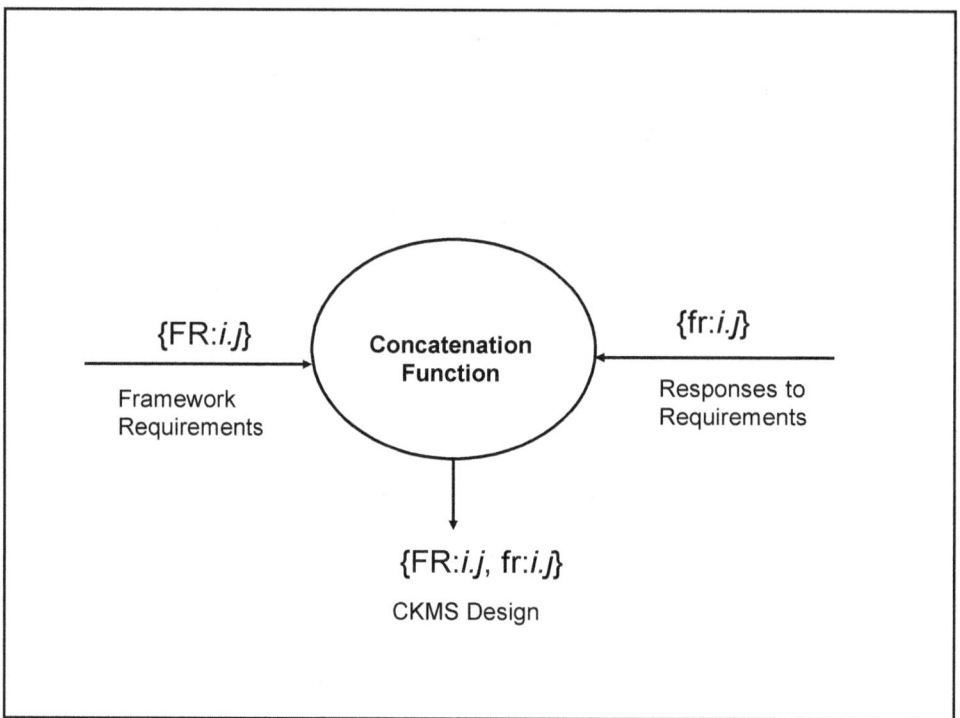

Figure 3: The CKMS Design Process for Framework Conformance

FR:2.4 The CKMS design **shall** specify a high-level overview of the CKMS system that includes:
  a) The use of each key type,
  b) Where and how the keys are generated,
  c) The metadata elements that are used in a trusted association with each key type,
  d) How keys and/or metadata are protected in storage at each entity where they reside,
  e) How keys and/or metadata are protected during distribution, and
  f) The types of entities to which keys and/or metadata can be delivered (e.g., user, user device, network device).

2.6 CKMS Profiles

A CKMS Profile specifies requirements that a qualifying CKMS, its implementation, and its operation must meet for a particular sector of organizations, such as Federal Agencies. A CKMS Profile specifies how the CKMS must be designed, implemented, tested, evaluated, and operated. A sector is a group of organizations that have common requirements for a product, system, or service. A CKMS Profile is a set of requirements concerning security and interoperability that must be satisfied by a CKMS as implemented in an operational system. This Framework may be used to derive a specific CKMS Profile for a particular sector. As with the Framework, one or more Profile Requirements correspond to each Profile Topic.

2.7 CKMS Framework and Derived Profile

Figure 4 depicts the relationship between the CKMS Framework and a derived sector Profile. When deriving a CKMS Profile from a Framework, the requirements of the Framework could be augmented and refined to meet the needs of the selected sector. For example, NIST could use this Framework to develop a Federal CKMS Profile for U.S. Federal Government agencies by selecting certain standards and protocols that comply with applicable Federal Information Processing Standards (FIPS), NIST Special Publications (SPs), and guidelines as necessary to meet the refined requirements.

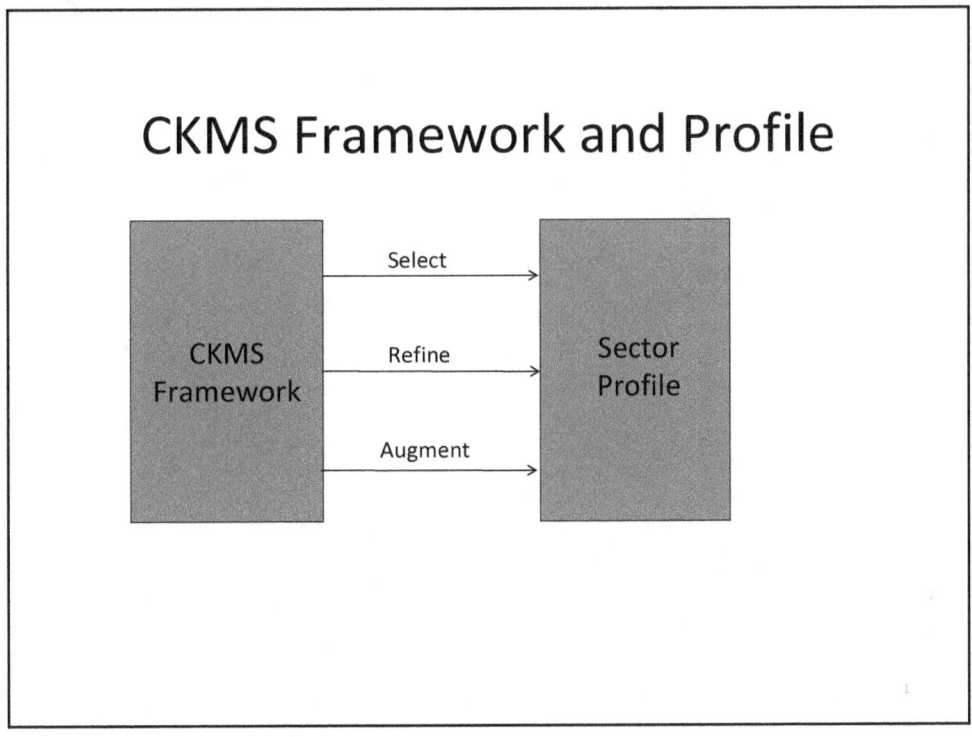

Figure 4: Relationship of CKMS Framework and Sector Profile(s)

2.8 Differences between a Framework and a Profile

A Framework requires that specific topics be addressed in the design of a CKMS, but it is not judgmental on the design itself. Any CKMS could be designed and specified in accordance with this Framework. On the other hand, a Profile states what requirements must be met in order to have a satisfactory CKMS for the designated using sector. A CKMS Profile makes judgments (i.e., specifies what is necessary to be implemented and used to be compliant with the Profile). CKMS that comply with this Framework may not comply with a particular profile. For example, **FR:2.1** in Section 2.1 requires that the CKMS design specify all cryptographic algorithms that are used by the CKMS. A U.S. Federal CKMS Profile might require that only NIST-approved cryptographic algorithms be used.

2.9 Example of a Distributed CKMS Supporting a Secure E-Mail Application

Figure 5 depicts a distributed CKMS that communicates among systems (shown in the figure as System A, System B, and System C) via the Internet. The CKMS consists of the union of all the CKMS modules (shown in the figure as CKMS Module A, CKMS Module B, and CKMS Module C). Each CKMS module is considered a logical entity within its system. Any parts of the system that perform CKMS functions are parts of the logical CKMS module at the time those functions are performed. In addition, parts of the CKMS module (e.g., an encryption algorithm) may be used by other applications (e.g., encrypting general data).

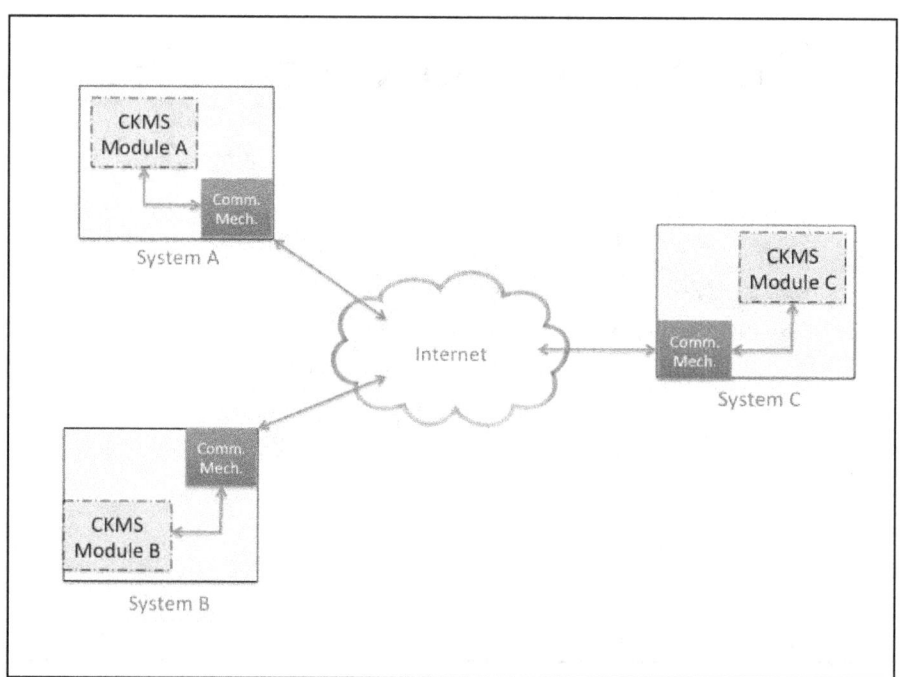

Figure 5: Example CKMS Overview

The actual communication mechanisms that interact with the other systems containing CKMS modules via the Internet are not part of the CKMS. However, the parts of protocols that perform CKMS functions (e.g., generating keys and providing key management information for insertion into the protocols) are considered part of the CKMS.

Figure 6 is an example of an email application that uses a distributed CKMS. The sender's email application interfaces with the CKMS module, which generates the keys that will be used to apply the required cryptographic protection for the email data to be sent to the intended receiver via the Internet and, if required, to apply cryptographic protection to the keys that will be transported to the receiving entity. The email application then hands off the protected key and the protected data to the communication mechanism for transmission. Note that the communication mechanism may also interact with the CKMS module as discussed for Figure 5.

The communication mechanism in the sender's system interacts with its CKMS module, as appropriate, prior to sending the cryptographically protected email to the email application. The email application sends the protected key to its local CKMS module to obtain the key that will subsequently be used to process the protected email data.

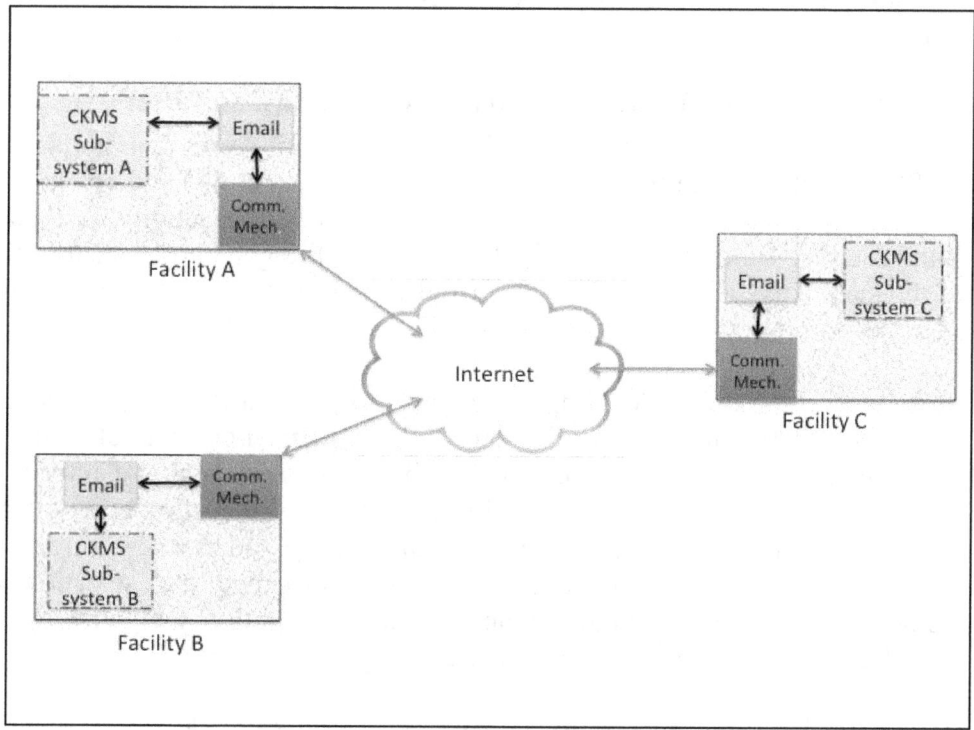

Figure 6: Example of a Secure Email Application

2.10 CKMS Framework Components and Devices

This CKMS Framework uses the term "component" to mean the hardware, software, and/or firmware required to construct the CKMS. The term "device" denotes a combination of components that serve a specific purpose. A CKMS can be as simple as a software program running on a single-user computer and supporting user applications. It can also be as complex as a variety of sub-systems, each containing many devices that provide key management services to numerous networked users and applications. A CKMS can be implemented in a single computer, or it may be widely distributed geographically and connected with a myriad of communications networks. Processors, communications media, storage units, etc. are all considered devices in this Framework.

A CKMS can be described as a set of policies, procedures, devices, and components that are designed to protect, manage, and establish cryptographic keys and metadata. The CKMS provides a set of functions that perform cryptographic key-management services on behalf of one or more organizations and their users. Collectively, these functions are presented as items for specification in a CKMS design (see Section 6.4).

FR:2.5 The CKMS design **shall** specify all major devices of the CKMS (e.g., the make, model, and version).

3. Goals

A CKMS should be designed to achieve specific goals. Some possible goals are discussed in this section.

3.1 Providing Key Management to Networks, Applications, and Users

There is an extensive use of cryptography in several security protocol standards (e.g., TLS, IKE, SSH, CMS), where both static keys (i.e., long-term keys) and ephemeral keys (i.e., keys used only for a single session or key management transaction) are used by the protocols themselves. While the focus of a CKMS is on the generation, distribution and storage of the static keys, a CKMS design must also include the generation, storage, and protection of the employed ephemeral keys as well.

The network over which the CKMS operates forms the communications backbone of the CKMS. The CKMS designer needs to understand the efficiency and reliability of the network so that the CKMS can be designed to have minimal negative impact on the network. The network size and scalability will provide some indication as to the number of users that the CKMS will need to handle both initially and in the future. Network characteristics, such as error properties, may also influence the selection of the cryptographic algorithms and cryptographic modes of operation that may extend (or worsen) the effects of communication errors after decryption is performed.

A CKMS can be built to serve a particular application (e.g., E-Mail, data storage, healthcare systems, and payment systems), or it can be designed to serve an entire enterprise, which encompasses many applications. A CKMS designed for a single application tends to be specifically designed for and closely integrated into the application, while an enterprise CKMS should be more generic so that common key-management functions may be shared as much as possible. A CKMS designer needs to have a good understanding of the application(s) that are to be supported, since they will likely affect the design choices.

The CKMS designer should also study the potential users of the system. How many users will use the CKMS and for what purposes? Are the users mobile or stationary? Do the users need to be knowledgeable about the CKMS, or will it be transparent to them? Are users operating under stressful conditions, where time is of the essence in performing their jobs? Some CKMS have failed because the designer assumed that the user understood the purpose and importance of cryptographic keys and public key certificates. If users are hampered from doing their work by a CKMS, then the CKMS will likely not be a successful security solution because it will not be used.

The goal of the CKMS designer is to specify a set of security mechanisms that function well together, provide a desired level of security that meets the needs of the application(s) and using organization(s), are affordable, and have a minimum negative impact on operations. These, as well as other CKMS goals, should be considered before a CKMS is designed, implemented, and operated.

FR: 3.1 The CKMS design **shall** specify its goals with respect to the communications networks on which it will function.

FR:3.2 The CKMS design **shall** specify the intended applications that it will support.

FR:3.3 The CKMS design **shall** list the intended number of users and the responsibilities that the CKMS places on those users.

3.2 Maximize the Use of COTS in a CKMS

Customers generally prefer Commercial Off-The-Shelf (COTS) products. Such products are often less costly to acquire, operate, and maintain than custom products designed and built for a single customer. However, COTS products designed and built to satisfy the "least common denominator" requirements of a large number of customers may not completely satisfy any of the customers. If the CKMS designer uses products that meet a range of requirements in a specific market sector, the CKMS will be more likely to be accepted in that market.

Using standard interfaces generally improves the extensibility of the product. Extensions and improvements should be allowed and supported by the COTS design of a CKMS so that the CKMS can be configured to meet varying functional and workload demands, including those based on the number of users, transactions, keys, and application data.

FR:3.4 The CKMS design **shall** specify the COTS products used in the CKMS.

FR:3.5 The CKMS design **shall** specify which security functions are performed by COTS products.

FR:3.6 The CKMS design **shall** specify how COTS products are configured and augmented to meet the CKMS goals.

3.3 Conformance to Standards

Much can be learned about a CKMS by examining the extent to which it utilizes applicable standards. Designs that comply with standards have the benefit of the experience and wisdom of those who developed the standards. In addition, if the standards have validation programs that measure compliance, there is increased confidence that the CKMS has been correctly implemented. See Appendix A for a list of appropriate standards with a brief description of each.

Standards specify how something shall or should be done. Multiple vendors can build to the same standard and, thereby, foster interoperability and competition. In addition, the use of standards often increases confidence in the product or implementation. There is increased confidence in a standard that was developed and reviewed by multiple parties working together. Complying with standards may also reduce the time-to-production for a product or the time-to-operation for an implementation, since the essential concepts do not have to be re-invented. Conformance-testing laboratories are useful in that errors in implementations may be found and eliminated before products are available in the marketplace.

FR:3.7 The CKMS design **shall** specify the Federal, national, and international standards that are utilized by the CKMS.

The availability of commercial products that conform to one or more standards in a CKMS architecture can greatly reduce the time and cost of producing a CKMS. The up-front cost of a conformance-tested product is likely to be more than offset by the saved costs of not having to adapt a non-conforming product or to build a similar product from scratch.

FR:3.8 For each standard utilized by the CKMS, the CKMS design **shall** specify which CKMS devices implement the standard.

FR:3.9 For each standard utilized by the CKMS, the CKMS design **shall** specify how conformance to the standard was validated (e.g., by a third party testing program).

3.4 Ease-of-use

Possibly the most significant constraint to the use of a CKMS is the difficulty that some systems present to the untrained user. Since most users are not cryptographic security experts, and security is often a secondary goal for them, the CKMS needs to be as transparent as possible. User interfaces that adapt to the expertise of the user can guide a new and less-trained user, while permitting an expert to use efficient short cuts and to bypass step-by-step guidance.

3.4.1 Accommodate User Ability and Preferences

Ease-of-use is very subjective. Something easy or obvious for one person may not be easy or obvious for another. Designers should keep in mind that users are not usually security experts, so they may not understand the purpose of the security feature that they are using. Since security is not usually the primary purpose of the product, transparent security is desirable. Negative experiences will likely affect the acceptance and use of a product. Therefore, a large segment of the potential user population needs to be satisfied that a security product is easy to use.

FR:3.10 The CKMS design **shall** specify all user interfaces to the system.

FR:3.11 The CKMS design **shall** specify the results of any user-acceptance tests that have been performed regarding the ease of using the proposed user interfaces.

3.4.2 Design Principles of the User Interface

While ease-of-use may be highly subjective and difficult to evaluate, several design principles for achieving this goal have been established. Ease-of-use design goals should assure that:

a) It is intuitive and easy to do the right thing using the CKMS. For example, key management function calls should be intuitively named.

b) It is difficult to do the wrong thing using the system. For example, the CKMS should not permit encryption using a signature-only key.

c) It is intuitive and easy to recover when a wrong thing is done. For example, the CKMS should provide an undo function that reverses the previous function.

This approach reduces the total lifecycle cost by reducing user support costs.

FR:3.12 The CKMS design **shall** specify the design principles of the user interface.

FR:3.13 The CKMS design **shall** specify all human error-prevention or failsafe features designed into the system.

3.5 Performance and Scalability

Performance improvements in computing and communications are major success stories in the computer industry. As performance improves, new applications require that even faster processing and communications be available. In the past, large key-distribution centers often serviced a maximum of several thousand security subscribers. Now, millions of people use the Internet regularly with ever-increasing demands, including new demands for security and for cryptographic keys. The need for secure processing, data storage and communications will continue to grow. This growth will require a CKMS to be scalable in order to meet the growing workload.

FR:3.14 The CKMS design **shall** specify the performance characteristics of the CKMS, including the average and peak workloads that can be handled for the types of functions and transactions implemented, and the response times for the types of functions and transactions under those respective workloads.

FR:3.15 The CKMS design **shall** specify the techniques that are supported and can be used to scale the system to increased workload demands.

FR:3.16 The CKMS design **shall** specify the extent to which the CKMS can be scaled to meet increased workload demands. This **shall** be expressed in terms of additional workload, response times for the workload, and cost.

4. Security Policies

A CKMS must be designed in a manner that supports the goals of each organization using the CKMS. Several types of policies will influence the design and use of a CKMS.

An organization may have different policies covering different applications or categories of information. For example, a military-related organization may have one set of policies covering classified information and a totally different set of policies covering personnel information.

An organization often creates and relies on layered policies, with high-level policies addressing issues at the information-management level and lower-level policies addressing specific rules for data-protection. A physical security policy may be specified in one document, and a communication security policy may be specified in another document. Computer systems are often built in accordance with their own computer security policy.

Layers of policies (e.g., information management, information security, physical security, computer security, communications security, and cryptographic key security) interrelate in many ways. Intermediate and lower layers of a policy hierarchy should provide more details on implementation and enforcement than the next higher layer. For example, an organizational Information Management Policy specifying that information must be protected against unauthorized disclosure should result in an Information Security Policy specifying the restriction of access to and use of the information only to properly identified and authorized people.

An organization may use a hierarchy of policies that will determine their requirements for a CKMS. Figure 7 provides an example of the policies that may be used and their relationships. Further discussion about these policies is provided in the following subsections.

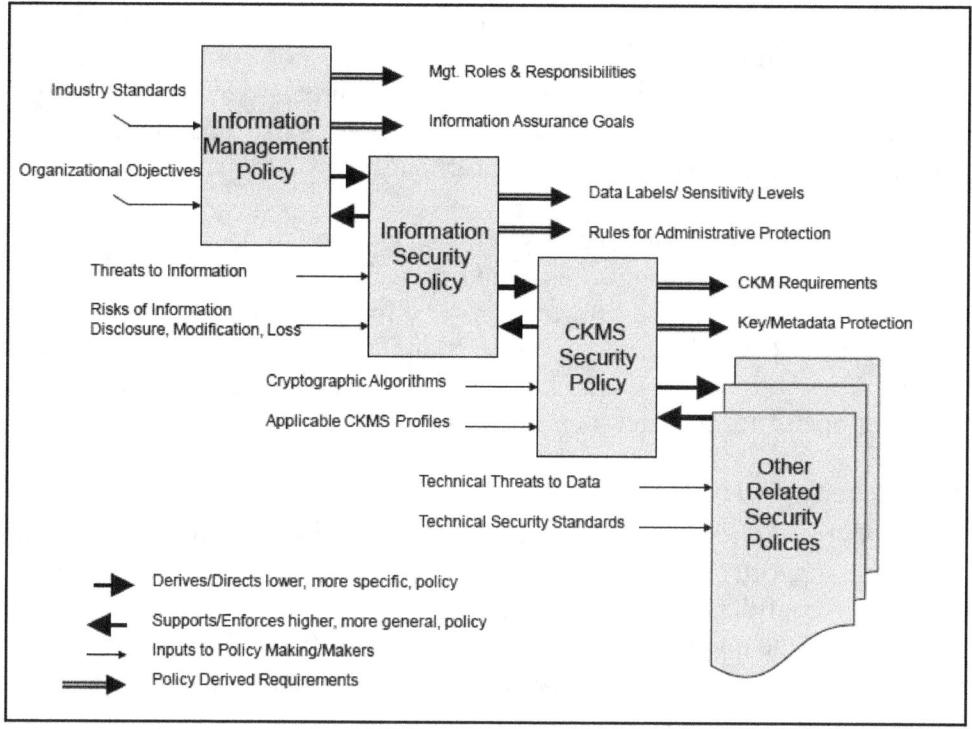

Figure 7: Related Security Policies

4.1 Information Management Policy

An organization's Information Management Policy specifies what information is to be collected or created, and how it is to be managed. An organization's management establishes this policy using industry standards of good practices, legal requirements regarding the organization's information, and organizational goals that must be achieved using the information that the organization will be collecting and creating.

An Information Management Policy typically identifies management roles and responsibilities and establishes the authorization required for people performing these information-management duties. It also specifies what information is to be considered valuable and sensitive and how it is to be protected. In particular, this highest policy layer specifies what categories of information need to be protected against unauthorized disclosure, modification or destruction. These specifications form the foundation for an Information Security Policy and dictate the levels of confidentiality, integrity, availability, and source-authentication protections that must be provided for differing categories of sensitive and valuable information.

4.2 Information Security Policy

An organization's Information Security Policy is created to support and enforce portions of the organization's Information Management Policy by specifying in more detail what information is to be protected from anticipated threats and how that protection is to be attained. The rules for collecting, protecting, and distributing valuable and sensitive information in both paper and electronic form are specified in this layer of policy. The inputs to the Information Security Policy include, but are not limited to, the Information Management Policy specifications, the potential threats to the security of the organization's information, and the risks involved with the unauthorized disclosure, modification, and destruction or loss of the information.

The outputs of the Information Security Policy layer include information sensitivity levels (e.g., low, medium, and high) assigned to various categories of information and high-level rules for protecting the information. The Information Security Policy may also be used to create a CKMS Security Policy that specifies the use and protection of cryptographic keys, algorithms, and mechanisms that provide confidentiality and integrity protection of the keys and their metadata for the organization.

4.3 CKMS Security Policy

The CKMS Security Policy specifies rules for the protection of keys and metadata that the CKMS must support. A CKMS Security Policy needs to establish and specify rules for protecting the confidentiality, integrity, availability, and source authentication of all cryptographic keys and metadata used by the CKMS. These rules cover the entire key lifecycle, including when they are operational, stored, and transported. The CKMS Security Policy may include the selection of all cryptographic mechanisms and cryptographic protocols that can be used by the CKMS. The CKMS Security Policy needs to be consistent with the higher-level policies of the organization. For example, if the Information Security Policy states that the confidentiality of electronically transmitted information is to be protected for up to 30 years, then the CKMS Security Policy and the CKMS design must be capable of supporting that policy.

The designer of a CKMS might not be a member of the organization that will be using the CKMS, and might not have access to the organization's policies, e.g., the organization may purchase a CKMS or the services of a CKMS that was developed external to the organization. The designer of the CKMS should create a set of security capabilities or features in the design that support the market for which the designer is creating the CKMS. These capabilities or features should be documented by the designer and can be considered to form the designer's initial CKMS Security Policy. The design documentation should state how and when the features are used to support the CKMS Security Policy. The organization may work with the designer or the CKMS service provider to develop a modified CKMS Security Policy, based on the initial CKMS Security Policy developed by the designer. Ultimately, it is the responsibility of the organizations that use the CKMS to assure that the CKMS design adequately supports, or can be configured to support, the (possibly modified) CKMS Security Policy.

The specific protections applied to each key type and its metadata (see Section 6) may be considered as supporting the Key Security Policy, which would be a part of the CKMS Security Policy. A Key Security Policy would state the policy for confidentiality, integrity and source authentication for the key and its metadata over the entire key lifecycle. These policies would then be supported by the CKMS.

A Key and Metadata Retention Policy specifying the length of time that keys and metadata are to be retained should also be part of the CKMS Security Policy. The Key and Metadata Retention Policy should be based on the sensitivity of the information that the keys and metadata protect. The CKMS should enforce the Key and Metadata Retention Policy. For example, the CKMS should protect keys and metadata throughout their security lifetimes, and then the CKMS should destroy the keys and metadata when they are no longer desired.

A CKMS Security Policy should be written so that the people responsible for maintaining the policy can easily understand the policy and correctly perform their roles and responsibilities. Note that security policies could be specified in a form (e.g., tables, formal specification languages, flow charts) that could be stored electronically and processed automatically within a CKMS. Policies specified in a formal language could be automatically enforced by a CKMS designed to do so. Such systems may be able to check themselves for proper functioning, diagnose current or potential problems, report the problem to the responsible organizational entity, and perhaps even automatically correct the problem.

FR:4.1 The CKMS design **shall** specify the CKMS Security Policy, including the configurable options and sub-policies that it is designed to enforce.

FR:4.2 The CKMS design **shall** specify how the CKMS Security Policy is to be enforced by the CKMS (e.g., the mechanisms used to provide the protection required by the policy).

FR:4.3 The CKMS design **shall** specify how any automated portions of the CKMS Security Policy are expressed in an unambiguous tabular form or a formal language (e.g., XML or ASN.1), such that an automated security system (e.g., table driven or syntax-directed software mechanisms) in the CKMS can enforce them.

4.4 Other Related Security Policies

A CKMS Security Policy may include or rely on other security policies. A CKMS design should state what other policies are required to be enforced for proper and secure operation of the CKMS. For example, a CKMS could be designed and implemented to provide all the physical protection and access control required to assure protection of the CKMS itself. It could also be designed assuming (and requiring) that external physical security and access control is provided by the facility in which the CKMS is installed and operated. Computer systems are often built to their own Computer Security Policy. An organization should create these policies in a logical structure that assigns roles for managing and enforcing the policies to appropriate parts of the organization.

FR:4.4 The CKMS design **shall** specify other related security policies that support the CKMS Security Policy.

4.5 Interrelationships among Policies

A CKMS designer should be aware of the various policies of organizations that may procure and use CKMS products or services. The designer could design a simple CKMS that enforces a simple key-management policy for a single organization or a complex CKMS product that can support a variety of security policies.

FR:4.5 The CKMS design **shall** specify the policies that are supported by the CKMS design and a summary of how they are supported by the design.

4.6 Personal Accountability

A policy of personal accountability requires that every person who accesses sensitive information be held accountable for his or her actions. Personal accountability may be a requirement in an Information Management Policy that results in specific features in the CKMS. A CKMS designer should determine if the CKMS is intended to support personal accountability. If it is, then mechanisms should be provided within the CKMS to support accountability for the management of keys and metadata.

FR:4.6 The CKMS design **shall** specify if and how personal accountability is supported by the CKMS.

4.7 Anonymity, Unlinkability, and Unobservability

An Information Management Policy may state that users of the secure information-processing system can be assured of anonymity, unlinkability, and unobservability. Anonymity assures that public data cannot be related to the owner. Unlinkability assures that two or more related events in an information-processing system cannot be related to each other. Finally, unobservability assures that an observer is unable to identify or infer the identities of the parties involved in a transaction.

FR:4.7 The CKMS design **shall** specify the anonymity, unlinkability, and unobservability policies that can be supported by the CKMS.

4.7.1   Anonymity

In order to provide privacy to entities, to adhere to applicable privacy laws, or to enhance security, a CKMS may require anonymity of CKMS transactions in terms of the entities that participate in the transaction. For privacy reasons, a CKMS may also require anonymity when associating keys and/or metadata with entities.

FR:4.8 The CKMS design **shall** specify which CKMS transactions have or can be provided with anonymity protection.

FR: 4.9 The CKMS design **shall** specify how CKMS transaction anonymity is achieved when anonymity assurance is provided.

4.7.2   Unlinkability

In order to provide privacy to entities, to adhere to applicable privacy laws, or to enhance security (by protecting against inferring who is associated with a given transaction), a CKMS may provide unlinkability protection for CKMS transactions in terms of the entities that participate in the transaction.

FR:4.10 The CKMS design **shall** specify which CKMS transactions have or can be provided with unlinkability protection.

FR:4.11 The CKMS design **shall** specify how CKMS transaction unlinkability is achieved.

4.7.3   Unobservability

In order to provide privacy to entities, to adhere to applicable privacy laws, or to enhance security (by protecting against inferring any information whose disclosure might not be desired), a CKMS may provide unobservability of CKMS transactions in terms of the entities that initiate or participate in the transaction.

FR:4.12 The CKMS design **shall** specify which CKMS transactions have or can be provided with unobservability protection.

FR:4.13 The CKMS design **shall** specify how CKMS transaction unobservability is achieved.

4.8 Laws, Rules, and Regulations

The security policies of an organization should conform to the laws, rules, and regulations of the locality, state, and nation(s) in which the CKMS will be used. If a CKMS is designed for international use, then it should be flexible enough to conform to national restrictions.

FR:4.14 The CKMS design **shall** specify the countries and/or regions of countries where it is intended for use and any legal restrictions that the CKMS is intended to enforce.

4.9 Security Domains

A *security domain* is a collection of entities, including their CKMS, in which each CKMS operates under the same security policy − known as the *Domain Security Policy*. When two mutually trusting entities are in the same security domain, the entities can exchange keys and metadata while providing the protections that are required by the Domain Security Policy.

When two entities are in different security domains, they may not be able to provide equivalent protection to the exchanged keys and metadata because they operate under different domain security policies. However, there are circumstances in which an entity in one domain can send keys and metadata to another entity in a different domain, even though the domain security policies are not completely identical.

An example of a security domain is a Public Key Infrastructure (PKI) that issues public key certificates (see [X.509]). The PKI operates under one or more documented certificate policies, and each public key certificate contains the certificate policies for which the certificate is valid. The relying entity (the certificate user) can examine the certificate and determine if the certificate provides acceptable security. However, when entities from different PKI domains wish to communicate, and hence use each other's certificates, the certificate policies of the two PKI domains should be examined and verified as offering equivalent security before the certificate should be used.

4.9.1   Conditions for Data Exchange

When an entity wishes to securely send a key and/or metadata to another entity, certain conditions must be satisfied:

a) There must be a means of sending and receiving the information, called a communications channel,

b) The two entities must have interoperable cryptographic capabilities (e.g., functionally identical encryption/decryption algorithms that utilize identical key lengths),

c) The two entities must subscribe to equivalent (though perhaps different) security policies, and

d) The two entities must trust each other (and perhaps other entities in the network) to enforce their own security policies[2].

If the entities belong to the same security domain, there is a high likelihood that each of these conditions can be met. But, if the entities do not belong to the same security domain, then these conditions are less likely to be satisfied. In the remainder of this section, assume that conditions a), b), and d) are met; the discussion will focus on condition c).

FR:4.15 The CKMS design **shall** specify design features that allow for the exchange of keys and metadata with entities in other security domains that are considered to offer equivalent but different security protections.

4.9.2   Assurance of Protection

Protection assurances include protecting a key and/or metadata from unauthorized disclosure, protecting the key and/or metadata from unauthorized modification, and knowing the source and destination of a key and/or metadata as required by the application. Suppose that entity A in security domain A wishes to send a cryptographic key and/or metadata to entity B in security domain B, and that conditions a), b), and d) above are satisfied. Suppose also that entity B wishes to receive the key and/or metadata and treat the received key and/or metadata exactly as it treats its own keys and/or

---

[2] An entity receiving data previously protected by one or more entities must trust the other entities to have properly enforced their own security policies.

metadata. That is to say, entity B in no way distinguishes the protections provided to the received key and/or metadata from those provided to its own keys and/or metadata. Before entity A sends the key and/or metadata, it must have assurance that the protection requirements in domain B's security policy are at least as good as those in domain A's security policy. Also, entity B would desire assurance that the protection requirements in domain A's security policy are at least as good as those in domain B's security policy. In essence, the two domains must have equivalent domain security policies

The protection assurances required for data sent from entity A to entity B are shown in Figure 8.

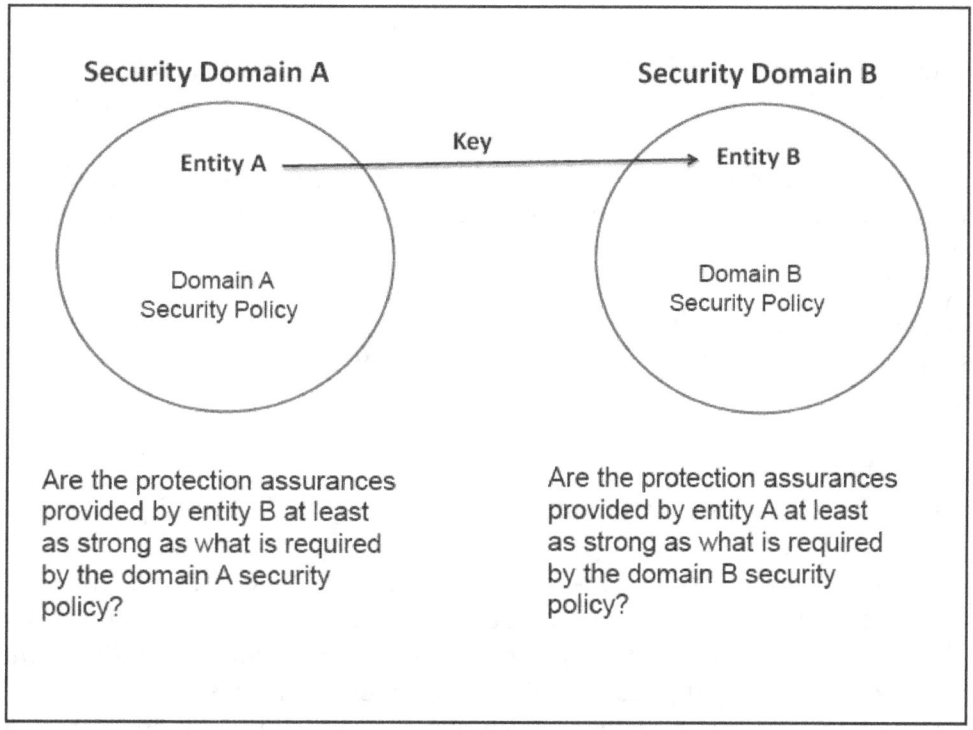

Figure 8: Protection Assurances between Security Domains

FR:4.16 The CKMS design **shall** specify the source and destination authentication policies that it enforces when sharing a key and/or metadata with entities in differing security domains.

FR:4.17 The CKMS design **shall** specify the confidentiality and integrity policies that it enforces when sharing a key and/or metadata with entities in differing security domains.

FR:4.18 The CKMS design **shall** specify what assurances it requires when communicating with entities from other security domains.

4.9.3   Equivalence of Domain Security Policies

Two security domains have equivalent security policies if the authority responsible for each security domain agrees to accept the other domain's policy as being equivalent to its own policy in terms of the security protections provided. The domain security policies have to be carefully examined before acceptance by the authorities responsible for each domain[3]. This process may be impossible if the authorities are not able to agree on the equivalence of protections. The authorities responsible for a security domain may restrict the security level of key and/or metadata, and therefore data, that they are willing to share with other domains in order to mitigate the consequences of any potential compromises. If entity A and entity B attempt to share a key and/or metadata, and security domain B has weaker policies than security domain A, then a sophisticated CKMS should, at a minimum, inform entity A of the possible security consequences.

If it is determined that the policies of the two domains are equivalent, an entity in one domain may share data with any entity in another equivalent domain, when appropriate.

FR:4.19 The CKMS design **shall** specify if and how it supports the review and verification of another domain's security before intra-domain communications are permitted.

FR: 4.20 The CKMS design **shall** specify how it detects, prevents or warns an entity of the possible security consequences of communicating with an entity in a security domain with weaker policies.

4.9.4   Third-Party Sharing

Suppose that entity A in security domain A and entity B in security domain B have equivalent domain security policies. In that case, it would be reasonable for entity A and entity B to share keys and/or metadata with any of the other members in either domain A or domain B, because each security domain has accepted the other domain's security policy. However, suppose that entity B also shares keys and metadata with a third entity, entity C in domain C. In this case, entity A and entity B have assurance that their respective domain security policies are equivalent and entity B and entity C have assurance that their respective domain security policies are equivalent. If entity B treats keys and/or metadata received from entity A in the same manner as its own keys and/or metadata, then entity A should expect that keys and/or metadata shared with entity B may also be shared with other equivalent security domains. When two entities examine each other's domain security policies for equivalence, they should pay close attention to each other's policies for sharing keys, metadata and other information with entities in other security domains.

---

[3] The process of determining the equivalence of security policies is similar to the Certificate Authority cross-certification process for Public Key Infrastructures.

4.9.5 Multi-level Security Domains

A security domain could contain entities, each of which supports the same multi-level Domain Security Policy. For example, the Domain Security Policy could provide either a high level or a low level of protection to the keys and/or metadata that it processes. In this case, the security domain acts much like two separate security domains because it must distinguish between the two levels of protection. Each entity must ensure that keys and/or metadata protected by the higher-level policy are always provided the higher level of protection, that keys and/or metadata protected by the lower-level policy cannot be confused with the higher-level keys and/or metadata, and that higher-level keys and/or metadata do not get confused with lower-level keys and/or metadata. This typically involves a multi-level operating system. See Figure 9. Physical entity B is divided into two logical entities: entity B_{HL} for high-level protection, and entity B_{LL} for low-level protection. The separation of the B_{HL} keys from the B_{LL} keys is maintained logically (as indicated by the dashed line in the figure) by the operating system. The advantage of a multi-level security domain is that it can process keys and/or metadata from entities operating at different security levels.

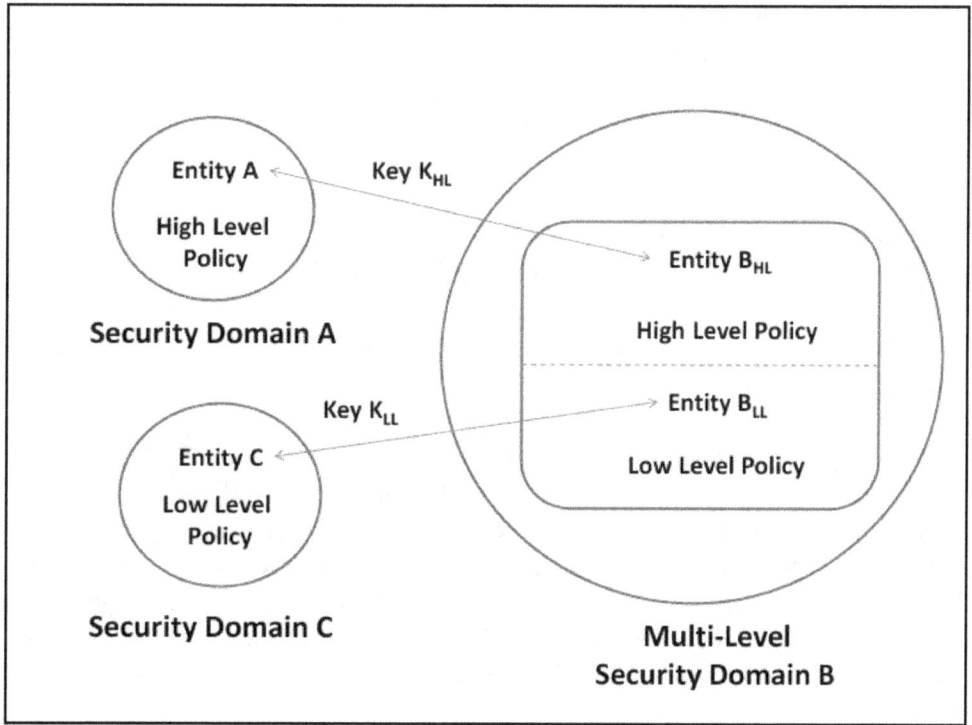

Figure 9: Multi-level Security Domain

FR: 4.21 The CKMS design **shall** specify whether or not it supports multilevel security domains.

FR:4.22 The CKMS design **shall** specify each level of security domain that it supports.

FR:4.23 If multilevel security domains are supported, the CKMS design **shall** specify how it maintains the separation of the keys and metadata belonging to each security level.

4.9.6 Upgrading and Downgrading

Under certain conditions, a domain authority may decide that a key and/or metadata from an entity in a lower-level security domain (a domain providing less protection) can be accepted and subsequently protected at the higher level required by its own Domain Security Policy. This process is called upgrading. Upgrading is not without risk and should only be done if the authority responsible for the higher-level domain has trust and confidence in the source and authenticity of the key and/or metadata from the lower level. A mistake in judgment by the domain authority could result in security compromises to the domain entities. Likewise, under certain conditions, the domain authority for a higher-level security domain may wish to pass a key and/or metadata down, or downgrade, to a lower-level domain entity. In this case, the domain authority for the higher-level domain should have confidence that the key and/or metadata being passed down only require the lower level of security provided by the receiver.

FR:4.24 The CKMS design **shall** specify if and how it supports the upgrading or downgrading of keys and metadata.

FR:4.25 The CKMS design **shall** specify how upgrading or downgrading capabilities are restricted to the domain authority.

4.9.7 Changing Domain Security Policies

From time to time, it may be desirable to modify or update a Domain Security Policy. The update may be the result of a management decision to upgrade the protections provided to the keys and metadata elements, it may be the result of a desire to be equivalent with another security domain, or it may be to support a new application.

Some CKMS may be designed so that their domain security policies may be configured to permit communications with entities in different domains. For example, a security domain may allow certain management officials to select the key and/or metadata management functions that are used to support various applications. These domains are said to be configurable. Even if a specific Domain Security Policy change is within the capability of a configurable system, the domain management personnel should still approve any policy change before the change is made.

FR:4.26 The CKMS design **shall** specify if and how its key and/or metadata management functions may be configured to support differing domain security policies and differing applications.

FR:4.27 The CKMS design **shall** specify if and how it can support changes in its Domain Security Policy by being reconfigured to accommodate communications with entities in different security domains.

5. Roles and Responsibilities

A CKMS may need to interface with humans that are performing specific management, user, and/or operational roles. Each role should have specific authorizations defined for it, and the persons performing that role should be provided access to a set of key and metadata management functions that are necessary for carrying out the responsibilities of the role. Examples of possible CKMS roles include, but are not limited to, the following:

a) **System Authority:** A system authority is responsible to executive-level management (e.g., the Chief Information Officer) for the overall operation and security of a CKMS. A system authority manages all operational CKMS roles. An operational role is a role that directly operates the CKMS.

b) **System Administrator:** System administrators are responsible for the personnel, daily operation, training, maintenance, and related management of a CKMS other than its keys. The system administrator is responsible for initially verifying individual identities and then establishing appropriate identifiers for all personnel involved in the operation and use of a CKMS. These include users, security auditors, cryptographic officers, key custodians, operators, maintenance workers, and agents required to vet the credentials of people seeking access to data in the system or use of the CKMS.

c) **Cryptographic Officer**: A cryptographic officer is authorized to perform cryptographic initialization and management functions on a CKMS and its cryptographic modules.

d) **Domain Authority**: A domain authority is responsible for defining and accepting a Domain Security Policy, for subsequently deciding the conditions necessary for communicating with other security domains, and then for assuring that the conditions are met.

e) **Key Custodian:** A key custodian is designated to distribute and/or load keys or key splits into a cryptographic module. Key custodians may be used to implement multi-party control and key splitting (See Section 6.7.4 and Section 6.7.5).

f) **Key Owner:** A key owner is an entity (e.g., person, group, organization, device, or cryptographic module) authorized to use a cryptographic key or key pair and whose identifier is associated with a cryptographic key or key pair. For public-private key pairs, the association is typically established through a registration process. A symmetric key may have a single specific owner, or multiple owners may share the key.

g) **CKMS User:** CKMS users utilize the CKMS when key management functions are required to support an application. CKMS users may be, and often are, key owners.

h) **Audit Administrator:** An audit administrator is responsible for auditing all aspects of a CKMS to verify its security and authorized operation. In particular, the audit administrator will manage and review the event log and should have no operational responsibilities for the CKMS. Audit administrators should not have access to any operational keys other than their own keys.

i) **Registration Agent:** A registration agent is responsible for registering new entities and binding their key(s) to their identifiers and perhaps other selected metadata. The registration agent may also enter entity information, keys, and metadata into a database used by the CKMS.

j) **Key-Recovery Agent:** A key-recovery agent is allowed to recover keys from backup or archive storage after identity verification and authorization of the requesting entity is performed in accordance with the CKMS Security Policy (see Sections 6.4.15 and 6.4.17).

k) **CKMS Operator:** A CKMS operator is authorized to operate (e.g., initiate the CKMS, monitor performance, and perform backups) a CKMS as directed by the system administrator.

Multiple individuals may be assigned to each role, and a single person may have multiple roles. However, certain roles should be separated so that no individual is assigned to both roles at the same time. For example, audit logs should be managed by someone other than a system administrator in order to detect administrative misuse or abuse. In addition, it is wise to rotate individuals from roles so as to minimize the likelihood of long-term abuses.

FR:5.1 The CKMS design **shall** specify each role employed by the CKMS, the responsibilities of each role, and how entities are assigned to each role.

FR:5.2 The CKMS design **shall** specify the key and metadata management functions (see Section 6.4) that can be used by entities fulfilling each role employed by the CKMS.

FR:5.3 The CKMS design **shall** specify which roles require role separation.

FR:5.4 The CKMS design **shall** specify how the role separation is maintained for the roles that require role separation.

FR:5.5 The CKMS design **shall** specify all automated provisions for identifying security violations, whether by individuals performing authorized roles (insiders) or by those with no authorized role (outsiders).

6. Cryptographic Keys and Metadata

6.1 Key Types

In general, cryptographic keys are categorized according to their properties and uses. Key properties may be *public*, *private*, or *symmetric*[4]. Keys may also have *static* (i.e., long term) or *ephemeral* (used only for a single session or key management transaction) properties. Key uses include *signature*, *authentication*, *encryption/decryption*, *key wrapping*, *RNG* (Random Number Generation), *master key*, *key transport*, *key agreement*, and *authorization*. [SP 800-57-part1] describes twenty different key types. Twenty-one key types are shown in Table 1 below (one compound key type in SP 800-57-part1 is divided into two simple key types in the table). Note that the italicized items in this paragraph are the actual terms that compose the key type names in the table. A CKMS may use these or other key types in its design.

Key Type
1) Private Signature Key
2) Public Signature Key
3) Symmetric Authentication Key
4) Private Authentication Key
5) Public Authentication Key
6) Symmetric Data Encryption/Decryption Key
7) Symmetric Key Wrapping Key
8) Symmetric RNG Key
9) Private RNG Key
10) Public RNG Key
11) Symmetric Master Key
12) Private Key Transport Key
13) Public Key Transport Key
14) Symmetric Key Agreement Key
15) Private Static Key Agreement Key
16) Public Static Key Agreement Key
17) Private Ephemeral Key Agreement Key
18) Public Ephemeral Key Agreement Key
19) Symmetric Authorization Key
20) Private Authorization Key
21) Public Authorization Key

Table 1: Key Types

FR: 6.1 The CKMS design **shall** specify and define each key type used.

---

[4] If it is not indicated in this document whether a key is asymmetric or symmetric, then either asymmetric or symmetric should be assumed.

6.2 Key Metadata

This section lists and describes the metadata that may be associated with keys. Key metadata is defined as information associated with a particular key that is explicitly recorded and managed by the CKMS. *In this Framework, the key associated with a particular set of metadata elements is referred to as "the key".*

The metadata that may be appropriate for a trusted association with a key should be selected by the CKMS designer, based upon a number of factors, including the key type, the key lifecycle state, and the CKMS Security Policy. A CKMS need not associate all applicable metadata with a given key, and a CKMS may not associate any metadata with some or all of the keys. See item u) in Section 6.2.1.

6.2.1 Metadata Elements

The following are typical metadata elements and their descriptions:

a) **Key Label:** A key label is a text string that provides a human-readable, and perhaps machine-readable, set of descriptors for the key. Examples of key labels include: "Root CA Private Key 2009-29" and "Maintenance Secret Key 2005."

b) **Key Identifier:** This element is used by the CKMS to select a specific key from a collection of keys. A key identifier is generally unique in a security domain. For public and private keys, a key identifier can be a hash value or portion of the hash value of the public key or can be assigned by the CKMS.

c) **Owner Identifier:** This element specifies the identifier (or identifiers) of the entity (or entities) that owns (or own) the key.

d) **Key Lifecycle State:** A key lifecycle state is one of a set of finite states that describe the current permitted conditions of a cryptographic key (see Section 6.3).

e) **Key Format Specifier:** This element is used to specify the format for the key. This can be accomplished by reference to the structure using object identifiers. For example, an RSA public key consists of the modulus and a public exponent. The format specifier should specify the sequence in which these two values are stored and the format in which each value is encoded. The Internet Engineering Task Force (IETF) has defined an object identifier for storing different forms of public keys, such as DSA, DH, RSA, EC, RSAPSS, and RSAOAEP keys. The object identifiers and related public key structures are defined in the following Internet RFCs: [RFC 3279], [RFC 4055], and [RFC 5480].

f) **Product used to create the Key:** This element specifies which cryptographic product was used to create or generate the key.

g) **Cryptographic Algorithm using the Key:** This element specifies the cryptographic algorithm that is intended to use the key. Examples include DSA, ECDSA, RSA, AES, TDEA, and HMAC-SHA1.

h) **Schemes or Modes of Operation:** This element defines the applicable schemes or modes of operation for performing a cryptographic function using a key. For asymmetric algorithms, it may specify the operation of discrete logarithm algorithms in a mathematical finite field, binary field, or Elliptic Curve (EC) field. For symmetric algorithms, this field may define the mode(s) of operation that can be used by the block cipher algorithm when using the key. Examples of modes of operation are Electronic Code Book (ECB), Cipher Block Chaining (CBC), Output Feedback Mode (OFB), and Counter with Cipher Block Chaining-Message Authentication Mode (CCM). For more information, see [SP 800-38A] through [SP 800-38F].

i) **Parameters for the Key:** This element specifies the parameters, if applicable, for a key. For example, a DSA key has the following domain parameters: large prime (p), small prime (q), and generator (g).

j) **Length of the Key:** This element specifies the length of the key in bits (or bytes). Examples include 2048 bits for an RSA modulus, and 256 bits for an elliptic curve key.

k) **Security Strength of the Key/Algorithm Pair:** This element is a number indicating the amount of work (that is, the base 2 logarithm of the number of operations) that is required to break (i.e., cryptanalyze) the cryptographic algorithm. For example, for a TDEA key of 168 bits (not including parity bits), the security strength is specified as 112 bits; for a 2048-bit RSA modulus, the security strength is specified as 112 bits. The security strength of a key/algorithm pair may be reduced if a previously unknown attack is discovered.

l) **Key Type[5]:** This element identifies the key type. Key types were discussed in Section 6.1.

m) **Appropriate Applications for the Key:** This element specifies applications for which the key may be used. Examples include Kerberos, Signed E-Mail, Trusted Time Stamp, Code Signing, File Encryption, and IPSEC.

n) **Key Security Policy Identifier:** This element identifies the security policy applicable to the key or key type. A Key Security Policy is a set of security controls that are used to protect the key or key type during the lifecycle of the key from generation to destruction (see Section 6.7 and [RFC 3647]). A Key Security Policy is typically represented by an object identifier registered by the CKMS organization. The Key Security Policy for individual keys or key types is part of, and should be consistent with, the CKMS Security Policy.

---

[5] Key type also implies key usage, since usage is one of the two factors that define key type. Thus, the key usage implied by the key type should be consistent with the application of the key.

o) **Key Access Control List (ACL)[6]:** An access control list identifies the entities that can access and/or use the keys as constrained by the key and metadata management functions (see Section 6.7). This Framework does not specify the access control list structure. The following are examples of such structures: a Microsoft Windows file/folder access control list consisting of zero or more access control entries, a Sun File System access control list, and while not a list, the Unix protection bits. In cases where interoperability is desired, the following items may require standardization: the syntax and semantics of the separators among access control entries, the ordering of entity and "access modes" within an access control entry, the entity identifier, and the designation of bits for different "access modes". If required for interoperability, these items should be included in an appropriately detailed design specification.

p) **Key Usage Count:** This element indicates the number of times that the key has been used.

q) **Parent Key:** This element points to the key from which the key associated with this metadata is derived. For example, a new key (i.e., the child key) could have been derived from a TLS master secret (i.e., the parent key) with its metadata.

This element may have two sub-elements:
  i. **Key Identifier:** The identifier for the parent key (see item b) above).
  ii. **Nature of the Relationship:** This element identifies how the parent key is related to the child key. An example of the relationship is a mathematical function that was used to create the child key using the parent key as one of the inputs. The relationship might be indicated by the identification of the mathematical function.

r) **Key Sensitivity:** This element specifies the sensitivity or importance of the key. It could relate to a risk level (e.g., Low, Moderate, or High) or a classification level (e.g., Confidential, Secret, or Top Secret)

s) **Key Protections[7]:** This element specifies the integrity, confidentiality, and source authentication protections applied to the key. A public key certificate is an example of key protection whereby the CA's digital signature provides both the integrity protection and source authentication (see [X.509]). A symmetric key and its hash value encrypted together is an example of confidentiality and integrity protection. When a key and its metadata are received from an external entity, the

---

[6] An ACL includes identifiers for authorized parties, their access mode or permission or authorization (such as create, initialize, use, entry, output, update, replace, revoke, delete, etc.), delegation rights for each access mode, and validity period for each access mode.

[7] A key can have multiple types of protection (e.g., integrity and confidentiality). The Framework permits the use of multiple cryptographic mechanisms for the same security service (e.g., digital signature and MAC for integrity).

protections should be verified before the key and metadata are operationally used. Generally, a single cryptographic function (e.g., HMAC or digital signature) is used to provide both integrity protection and source authentication.

This element may have several sub-elements:
   i. The mechanism used for integrity protection (e.g., hash value, MAC, or digital signature),
   ii. The mechanism used for confidentiality protection (e.g., key wrapping or key transport),
   iii. The mechanism used for source authentication (e.g., MAC or digital signature), and
   iv. An indication of the protections that are enforced by a particular non-cryptographic trusted process.

t) **Metadata Protections (can be a subset of the key protections or can be different):** This element specifies the mechanisms used to provide integrity, confidentiality, and source authentication to the associated metadata. Generally, the same mechanism will be used to protect the key and its metadata, especially if the key and metadata are transmitted or stored together.

This element may have several sub-elements:
   i. The mechanism used for integrity protection (e.g., hash value, MAC, or digital signature),
   ii. The mechanism used for confidentiality protection (e.g., encryption),
   iii. The mechanism used for source authentication, and
   iv. An indication of the protections that are enforced by a particular non-cryptographic trusted process.

u) **Trusted Association Protections (i.e., how the trusted association of metadata to the key is protected) (can be part of key protection in item s) above):** This information is implicitly provided if the key and metadata are protected as one aggregated item using the protection listed in item s) above. Otherwise, the following should be provided for each trusted association protection:
   i. The mechanism used for integrity protection (e.g., hash value, MAC, digital signature, or trusted process), and
   ii. The mechanism used for source authentication (e.g., cryptographic mechanism or non-cryptographic trusted process).

v) **Date-Times:** There are several important date-times for the lifecycle state transitions of a key:
   i. The generation date: The date-time that a key was generated,
   ii. The association date: The date-time that a key was associated with its metadata for the first time,
   iii. The activation date: The date-time that a key was first used,
   iv. The future activation date: The date-time that a key is first to be used,

v.   The renewal date: The date-time that a public key was renewed and allowed to be used for a longer period of time, e.g., by generating a new certificate for the same public key as was provided in an old certificate (see Section 6.4.7),

vi. The future renewal data: The date-time that a public key is to be renewed and allowed to be used for a longer period of time (e.g., by generating a new certificate for the same public key as was provided in an old certificate),

vii. The date of the last rekey: The date-time that a key was replaced with a new key that was generated so that it is completely independent of the key that was replaced,

viii. The future rekey date: The date-time that the key is to be replaced with a new key that will be generated so that it is completely independent of the key being replaced,

ix. The date of the last usage of the key: The date-time that the key was last used.

x.   The deactivation date: The date-time that a key was deactivated,

xi. The future deactivation date: The date-time that a key is to be deactivated,

xii. The expiration date: The date-time that a key's useful lifetime was terminated permanently,

xiii. The revocation date: The date-time after which a key was no longer considered valid,

xiv. The compromise date: The date-time that a key was known or suspected to have been compromised and was marked for replacement and not renewal,

xv. The destruction date: The date-time that a key was destroyed, and

xvi. The future destruction date: The date-time that a key is to be destroyed.

w) **Revocation Reason:** If a key is revoked, this element specifies the reason for the revocation. Examples include a compromise due to an adversary having the key, a compromise due to an adversary having the cryptographic module containing the key, a loss of the key, a loss of the cryptographic module containing the key, a suspected key compromise, the key owner left the sponsoring organization, and a key misuse by the owner.

The dates and times used in the above listed metadata elements, as well as various CKMS transaction dates and times, may be required to be both accurate and from an authoritative source, such as a Network Time Protocol (NTP) server. In addition, some of the transactions may require time stamps from a trusted third-party. Trusted third-party time stamping is described in [RFC 3161] and [SP 800-102].

FR:6.2 For each key type used in the system, the CKMS design **shall** specify all metadata elements selected for a trusted association, the circumstances under which the metadata elements are created and associated with the key, and the method of association (i.e., cryptographic mechanism or trusted process).

FR: 6.3 For each cryptographic mechanism used in the Key Protections metadata element (item s above), the CKMS design **shall** specify the following:
  i. The cryptographic algorithm: See item g) above.
  ii. The parameters for the key: See item i) above.
  iii. The key identifier: See item b) above.
  iv. The protection value: This element contains the protection value for integrity protection, confidentiality protection, or source authentication. For example, a properly implemented MAC or digital signature technique may provide for integrity protection and/or source authentication.
  v. When the protection was applied.
  vi. When the protection was verified.

FR:6.4 For each non-cryptographic trusted process used in the Key Protections metadata element (item s above), the CKMS design **shall** specify the following:
  i. The identifier of the process used to distinguish it from other processes, and
  ii. A description of the process or a pointer to a description of the process.

FR:6.5 For each cryptographic mechanism used in the Metadata Protections metadata element (item t above), the CKMS design **shall** specify the following:
  i. The cryptographic algorithm.
  ii. The parameters for the key.
  iii. The key identifier.
  iv. The protection value (e.g., MAC, digital signature).
  v. When the protection was applied.
  vi. When the protection was verified.

Generally, the same mechanism will be used for the key and bound metadata, especially if the key and metadata are bundled together.

FR:6.6 For each non-cryptographic trusted process used in the Metadata Protections metadata element (item t above), the CKMS design **shall** specify the following:
  i. The identifier that is used to distinguish this process from other processes, and
  ii. A description of the process or a pointer to a description of the process.

FR:6.7 For each cryptographic mechanism used in the Trusted Association Protections metadata element (item u above), the CKMS design **shall** specify the following:
  i. The cryptographic algorithm,
  ii. The parameters for the key,
  iii. The key identifier,
  iv. The protection value (e.g., MAC, digital signature),
  v. When the protection was applied, and

vi. When the protection was verified.

FR:6.8 For each non-cryptographic trusted process used in the Trusted Association Protections metadata element (item u above), the CKMS design **shall** specify the following:
   i.   The identifier that is used to distinguish this process from other processes, and
   ii. A description of the process or a pointer to a description of the process.

FR:6.9 The CKMS design **shall** specify the accuracy and precision required for dates and times used by the system.

FR:6.10 The CKMS design **shall** specify what authoritative time sources are used to achieve the required accuracy.

FR:6.11 The CKMS design **shall** specify how authoritative time sources are used to achieve the required accuracy[8].

FR:6.12 The CKMS design **shall** specify which dates, times, and functions require a trusted third-party time stamp.

6.2.2   Required Key and Metadata Information

A CKMS design needs to make certain information clear regarding how keys and metadata are managed.

FR:6.13 For each key type, the CKMS design **shall** specify the following information regarding keys and metadata elements:
   a) The key type
   b) The cryptoperiod (for static keys)
   c) The method of generation
      i.   The RNG used
      ii. A key generation specification (e.g., [FIPS 186] for signature keys, [SP 800-56A] for Diffie-Hellman key establishment keys)
   d) For each metadata element, include
      i.   The source of the metadata
      ii. How the metadata is vetted
   e) The method of key establishment
      i.   The key transport scheme (if used)
      ii. The key agreement scheme (if used)
      iii. The protocol name (if a named protocol is used)
   f) The disclosure protections (e.g., key confidentiality, physical security)
   g) The modification protections (e.g., a MAC or a digital signature)

---

[8] For example, the use of an NTP server and an NTP protocol to synchronize with the authoritative time source.

h) The applications that may use the key (e.g., TLS, EFS, S/MIME, IPSec, PKINIT, SSH, etc.)
i) The applications that are not permitted to use the key
j) The key assurances
   i. Symmetric key assurances (e.g., format checks)
     • Who obtains the assurance
     • The circumstances under which it is obtained
     • How the assurance is obtained
   ii. Asymmetric key assurances (e.g., assurance of possession and validity)
     • Who obtains the assurances
     • The circumstances under which the assurance is obtained
     • How the assurance is obtained
   iii. Domain parameter validity checks
     • Who performs the validity check
     • The circumstances under which the checking is performed
     • How the assurance of domain parameter validity was obtained.

FR:6.14 The CKMS design **shall** specify all syntax, semantics, and formats of all key types and their metadata that will be created, stored, transmitted, processed, and otherwise managed by the CKMS.

6.3 Key Lifecycle States and Transitions

A key may pass through several states between its generation and its destruction. This section is based on Section 7, Key States and Transitions, from [SP 800-57-part1]. Possible states of a key include: Pre-Activation, Active, Deactivated, Compromised, Destroyed, Destroyed Compromised, and Revoked. Note that the CKMS designer selects and defines the key states and transitions that are appropriate for the CKMS and its likely applications.

FR:6.15 The CKMS design **shall** specify all the states that the CKMS keys can attain.

FR:6.16 The CKMS design **shall** specify all transitions between the CKMS key states and the data (inputs and outputs) involved in making the transitions.

6.4 Key and Metadata Management Functions

The key and metadata management functions described in this section are performed by the CKMS on keys or metadata for management purposes. The authentication and authorization of the calling entities is performed by an Access Control System (ACS), as described in Section 6.7.

A CKMS should provide for the creation, modification, replacement, and destruction of keys and their metadata. Depending on the function, the input and/or output may have integrity, source authentication, and/or confidentiality services applied to them.

In the case of an input to a function, the function may need to process protections placed on the input by another entity. For example, for the key-entry function, the entity providing the key (i.e., the key source[9]) may have digitally signed the plaintext key and then encrypted the signed result. Therefore, for this example, the key-entry function will need to decrypt the input and perform digital signature verification to authenticate the key source and verify the integrity of the key.

In the case of an output from a function, the function may need to apply security services. For example, for the key-output function, the invoker of the function may desire to output a key that is encrypted and then digitally signed. The key-output function would then apply encryption and digital signature generation to the key as appropriate for the intended recipient.

FR:6.17 The CKMS design **shall** specify the key and metadata management functions to be implemented and supported.

FR:6.18 The CKMS design **shall** identify the integrity, confidentiality, and source-authentication services that are applied to each key and metadata management function parameter implemented in the CKMS.

6.4.1   Generate Key

When a user requires a key, the user should request that the key be generated by the CKMS. The user may need to specify the type of key and other necessary parameters (e.g., the name of the key-generation technique), including some metadata that needs to be associated with the key when requesting this function. The function may return a key identifier that is a pointer to the key and perhaps its metadata. If the user wishes to actually know the key value, then the key-output function (see Section 6.4.20) could be used in some circumstances.

Key-generation techniques typically depend on the specifications of the cryptographic algorithm paired with the key (see [FIPS 186]). Different algorithms make use of keys conforming to differing specifications (e.g., lengths and formats). Key generation for asymmetric algorithms involves the generation of a key pair. The generation of keys requires the use of a random number generator that is designed for cryptographic purposes. For example, NIST has published several approved random number generators (see [SP 800-90A]) and instructions on key generation (see [FIPS 186]).

The key-generation function may also provide for the selection or input of metadata that is associated with the generated key.

FR:6.19 The CKMS design **shall** specify the key-generation methods to be used in the CKMS for each type of key.

---

[9] The source of the key may or may not be the entity using the key-entry function.

FR:6.20 The CKMS design **shall** specify the underlying random number generators that are used to generate symmetric and private keys.

6.4.2 Register Owner

The initial registration of a security entity (i.e., individual (person), organization, device or process) and a cryptographic key with metadata is a fundamental requirement of every CKMS. This requirement is difficult to fully automate while preserving security (i.e., protecting from an impersonation threat) and thus, it usually requires human interactions. There typically exists a registration process in a CKMS that binds each entity's initial set of secret, public, or private keys with the entity's identifier and perhaps other metadata. The process of binding an owner's identifiers and keys involves either an initial identity proofing of the owners or relying on the pre-existing identity of the owner in the CKMS.

FR:6.21 The CKMS design **shall** specify all the processes involved in owner registration, including the process for binding keys with the owner's identifier.

6.4.3 Activate Key

The activation function provides for the transition of a cryptographic key from the pre-activation state to the active state. This function may automatically activate the key immediately after generation. Alternatively, this function may generate a date-time metadata value that indicates when the key becomes active and can be used. A deactivation date-time may also be established using this function.

FR:6.22 The CKMS design **shall** specify how each key type is activated and the circumstances for activating the key.

FR:6.23 For each key type, the CKMS design **shall** specify requirements for the notification of key activation, including which parties are notified, how they are notified, what security services are applied to the notification, and the time-frames for notification(s)[10].

6.4.4 Deactivate Key

This function transitions a key into the deactivated state. A cryptographic key is generally given a deactivation date and time when it is created and distributed. In some instances, deactivation may also be based on the number of uses or the amount of data protected. This deactivation information may be associated with the key as metadata. The period of time between activation and deactivation is generally considered the cryptoperiod of a key. This time usually has a maximum value based, in part, on the sensitivity levels of the data it is protecting and the threats that could be brought against the CKMS (see [SP 800-57-part1] for further discussion). The cryptoperiod can be shortened, based on the concerns of the cryptographic officer in charge of the key and data. The CKMS Security

---

[10] For example, notification could be once immediately before activation, or every *n* units of time until activation, starting at some time in advance, or with increasing frequency as the activation time approaches.

Policy should state the maximum allowable cryptoperiod of any key type used to protect the data covered by the policy.

FR: 6.24 The CKMS design **shall** specify for each key type how deactivation of the key is determined (e.g., by cryptoperiod, by number of uses, or by amount of data).

FR: 6.25 The CKMS design **shall** specify how each key type is deactivated (e.g., manually or automatically, based on the deactivation date-time, the number of usages, or the amount of protected data).

FR: 6.26 The CKMS design **shall** specify how the deactivation date-time for each key type can be changed[11].

FR:6.27 For each key type, the CKMS design **shall** specify requirements for advance notification of the deactivation of the key type, including which CKMS supported roles are notified, how they are notified, what security services are applied to the notification, and the time-frames for notification(s).

6.4.5   Revoke Key

Key revocation is used in cases where the authorized use of a key needs to be terminated prior to the established cryptoperiod of that key. A cryptographic key should be revoked as soon as feasible after it is no longer authorized for use (e.g., the key has been compromised). Revoking a key includes marking the key as no longer authorized for use to apply cryptographic protection or to process already protected information. Security entities that have been, that are, or that will be using the key (i.e., relying parties) need to be notified that the key has been revoked. This may involve the publication of a revocation list identifying keys that have been revoked. Other forms of revocation notification may be supported in key-management systems.

FR:6.28 The CKMS design **shall** specify when, how, and under what circumstances revocation is performed and revocation information is made available to the relying parties.

6.4.6   Suspend and Re-Activate a Key

A key may be temporarily suspended and later re-activated[12]. Examples of situations that may warrant suspension, as opposed to irreversible revocation, include: the owner is not available for an extended period of time, the key has been misused, a possible compromise is under investigation, or a token containing a key has been misplaced. In

---

[11] For example, over time, the advancements in key exhaustion technology may improve at a faster rate than expected, or new attacks that lower the bits of security strength provided by the key and its algorithm may be discovered. Thus, the key-deactivation date may require modification.

[12] Suspension is a temporary deactivation. In other words, while deactivation is generally irreversible, suspension can be reversed in order to re-activate the key.

addition to a security-issue-related revocation (since suspension is nothing but revocation, albeit reversible), the security of re-activating a suspended key is also critical.

If a suspension is to apply to remote entities holding the key, as well as the local calling entity, then provisions must be made for notifying the other entities of the suspension and also the re-activation.

FR:6.29 The CKMS design **shall** specify how, and under what circumstances, a key is suspended.

FR:6.30 The CKMS design **shall** specify how suspension information is made available to the relying or communicating parties.

FR:6.31 The CKMS design **shall** specify how, and under what circumstances, a suspended key is re-activated.

FR:6.32 The CKMS design **shall** specify how the suspended key is prevented from performing security services.

FR:6.33 The CKMS design **shall** specify how re-activation information is made available to the relying or communicating parties.

6.4.7   Renew a Public Key

Public key certificates contain a public key of an asymmetric key pair (i.e., the subject key) and a validity period for that certificate. It may be desirable to have a public key validity period that is shorter than the subject key's cryptoperiod. This reduces the size of revocation lists and revocation information, but requires certificates to be issued more frequently. Renewal establishes a new validity period for an existing subject public key beyond its previous validity period by issuing a new certificate containing the same public key with a new validity period. The sum of the renewal periods for a given public key must not exceed the cryptoperiod of the key.

Advance notification is useful for continuity of operations and mission so that the appropriate set of new keys and associated metadata can be issued to appropriate parties. For example, upon the expiration of an entity's current public key certificate, the entity may need to request either the renewal of the existing public key or the establishment of a new public key.

FR:6.34 The CKMS design **shall** specify how and the conditions under which a public key can be renewed.

FR:6.35 For each key type, the CKMS design **shall** specify requirements for advance notification of the renewal of the key type, including which parties are notified, how they are notified, what security services are applied to the notification, and the time-frames for notification(s).

6.4.8   Key Derivation or Key Update

When a key is derived from other information, some of which is secret, in a non-reversible manner, the process is called key derivation. Key derivation is often used in key establishment protocols to derive a shared key from a common shared secret (see [SP 800-56A], [SP 800-56B], [SP 800-56C], and [SP 800-135]).

Key derivation may also be used to derive a key from another key (see [SP 800-108]) or from a password (see [SP 800-132]). In the case where a key (e.g., K_1) is used to derive another key (K_2), and the derived key (K_2) is used to **replace** the original key (i.e., K_1), then the process is called key update. In the past, keys were merely updated in order to avoid having to use a key establishment protocol to establish a new key; all entities sharing the key merely updated the key to form a new key without using any other secret data. This process of key updating has the possible security exposure that an adversary who obtains a key (by compromise or cryptanalysis) and knows the update transformation can update the known key to any of its future updates.

FR:6.36 The CKMS design **shall** specify all processes used to derive or update keys and the circumstances under which the keys are derived or updated.

FR:6.37 For each key type, the CKMS design **shall** specify requirements for advance notification of the derivation or update of the keys, including which parties are notified, how they are notified, what security services are applied to the notification, and the time-frames for notification(s).

6.4.9   Destroy Key and Metadata

Keys and some portion of their metadata should be destroyed beyond recovery when they are no longer to be used. Destroying a key in a high-security application can be a complex process, depending on the storage media for the key and the extent of distribution of key copies. Historically, the secure burning of paper keying material (paper tape, punched cards, or printed key lists) in a prescribed manner was used. Keys in electronic storage media may be overwritten with random patterns of zeros and ones. Magnetic media that has a propensity for retaining low levels of magnetism may be physically destroyed, degaussed, or over-written with various bit patterns numerous times. Designers should include provisions for destroying a key in backup storage media if such media are utilized.

FR:6.38 The CKMS design **shall** specify how and the circumstances under which keys are intentionally destroyed and whether the destruction is local to a component or universal throughout the CKMS.

FR:6.39 For each key type, the CKMS design **shall** specify requirements for an advance notification of key destruction, including which parties are notified, how they are notified, what security services are applied to the notification, and the time-frames for notification(s).

6.4.10 Associate a Key with its Metadata

A cryptographic key may have several metadata elements associated with it. The CKMS designer must determine which metadata must or can be associated with a key and also the protection mechanism that provides the association. Depending on the nature of the information stored in a metadata element, the metadata element may require confidentiality protection, integrity protection, and source authentication. The association function uses cryptography or a trusted process to provide this protection.

FR: 6.40 For each key type used, the CKMS design **shall** specify what metadata is associated with the key, how the metadata is associated with the key, and the circumstances under which metadata is associated with the key.

FR: 6.41 For each key type used, the CKMS design **shall** describe how the following security services (protections) are applied to the associated metadata: source authentication, integrity, and confidentiality.

6.4.11 Modify Metadata

The modify metadata function can be used to modify existing writable metadata that is associated with a key. Metadata that has been associated with a key should not be modifiable by an unauthorized entity. For example, if the identifier of the key's owner is included in the metadata, an unauthorized entity should not be permitted to modify the key owner identifier or add additional owners. The binding of a key to its metadata can be achieved using a MAC or a digital signature. The integrity of the key and its metadata may be determined by verifying the MAC or digital signature.

FR: 6.42 The CKMS design **shall** specify the circumstances under which associated metadata is modified.

6.4.12 Delete Metadata

This function deletes metadata (for which delete permission has been granted) associated with a key. Metadata elements may be deleted as an entire complete group, as individual elements, or as a specific subset of the elements.

FR: 6.43 The CKMS design **shall** specify the circumstances under which the metadata associated with a key is deleted.

FR: 6.44 The CKMS design **shall** specify the technique used to delete associated metadata.

6.4.13 List Key Metadata

This function allows an entity to list the metadata elements of a key for which the entity is authorized. An entity may have multiple keys with associated metadata in storage. There may be keys for digital signature generation and verification, authentication, encryption/decryption, data integrity, key establishment, and key storage. Authorization to use a key does not automatically imply access to every metadata element associated with the key, but it may be impractical to remember all the values of every metadata element associated with a key. Therefore, the list metadata function may be very useful.

FR:6.45 For each key type, the CKMS design **shall** specify which metadata can be listed by authorized entities.

6.4.14 Store Operational Key and Metadata

Operational key and metadata storage involves the moving of keys and/or metadata to a medium from which the stored data may later be recovered. Keys and metadata should be physically or cryptographically protected when stored outside of a cryptographic module (see [SP 800-57-part1]).

FR:6.46 For each key type, the CKMS design **shall** specify: the circumstances under which keys of each type and their metadata are stored, where the keys and metadata are stored, and how the keys and metadata are protected.

6.4.15 Backup of a Key and its Metadata

Key and metadata backup involves the copying of keys and/or metadata to a safe facility so that it can be recovered if the original (operational) copy is lost, modified, or otherwise unavailable. Backup copies of keys and metadata may be located in the same or a different facility than the operational keys/metadata to assure that the keys and metadata can be recovered when needed, even after a natural or man-made disaster. Keys/metadata may be backed-up by the owner or a trusted entity.

FR:6.47 The CKMS design **shall** specify how, where, and the circumstances under which keys and their metadata are backed up.

FR:6.48 The CKMS design **shall** specify the security policy for the protection of backed-up keys/metadata[13].

FR:6.49 The CKMS design **shall** specify how the security policy is implemented during the key and metadata back-up, e.g., how the confidentiality and multi-party control requirements are implemented during transport and storage of the backed-up keys and metadata.

6.4.16 Archive Key and/or Metadata

The archive of keys and/or metadata involves placing keys and/or metadata in a safe, long-term storage facility so that they can be recovered when needed. The archive supports the Key and Metadata Retention Policy (see Section 4.3). Archived keys and/or metadata must be physically or cryptographically protected. Keys used to protect the keys and/or metadata in an archive are called archive keys. These archive keys will also have cryptoperiods, and the continued protection provided to the archived keys and/or metadata needs to be considered when the cryptoperiod of the archive key expires. This

---

[13] For example, two-person control might be required.

may include physical protection and/or the generation of a new archive key for the same or a stronger cryptographic algorithm, and re-encryption of the archived keys and/or metadata under the new archive key.

Key and metadata archiving usually requires provisions for moving archived keys and/or metadata to new storage media when the old media are no longer readable because of the aging of, or technical changes to, the media and media readers. Archived keys and/or metadata should be recovered from the old storage medium and stored on the new storage medium; the keys should be destroyed on the old storage medium after the transfer. When performing key and/or metadata archival or destruction, applicable laws and regulations must be considered so that the keys and/or metadata are available for the required period of time.

FR:6.50 The CKMS design **shall** specify how, where, and the circumstances under which keys and/or their metadata are archived.

FR:6.51 The CKMS design **shall** specify the technique for the secure destruction of the key and/or metadata or the secure destruction of the old storage medium after being written onto a new storage medium.

FR:6.52 The CKMS design **shall** specify how keys and/or their metadata are protected after the cryptoperiod of an archive key expires.

6.4.17 Recover Key and/or Metadata

Key and/or metadata recovery involves obtaining a copy of a key and/or its metadata that has been previously backed-up, archived, or stored. The key and/or metadata can be recovered by an authorized entity (e.g., its owner or by a trusted entity) after all the rules for recovery have been fulfilled and verified. The CKMS Security Policy should state the conditions under which a key and/or metadata may be recovered.

FR: 6.53 The CKMS design **shall** specify the CKMS recovery policy for keys and/or metadata.

FR:6.54 The CKMS design **shall** specify the mechanisms used to implement and enforce the recovery policy for keys and/or metadata.

FR:6.55 The CKMS design **shall** specify how, and the circumstances under which, keys and/or metadata are recovered from each key database or metadata storage facility.

FR: 6.56 The CKMS design **shall** specify how keys and/or metadata are protected during recovery.

6.4.18 Establish Key

Key establishment is the process by which a key is securely shared between two or more entities. The key may be transported from one entity to another (key transport), or the key may be derived from information shared by the entities (key agreement). The method of transporting keys or sharing information may be either manual (e.g., sent by courier) or automated (e.g., sent over the Internet).

FR: 6.57 The CKMS design **shall** specify how, and the circumstances under which, keys and their metadata are established.

6.4.19 Enter a Key and Associated Metadata into a Cryptographic Module

The key entry function is used to enter one or more keys and associated metadata into a cryptographic module in preparation for active use. Keys and metadata may be entered in plaintext form, in encrypted form, as key splits, in an integrity-protected form (e.g., in a signed certificate) or any combination thereof.

FR: 6.58 The CKMS design **shall** specify how, and the circumstances under which, keys and metadata are entered into a cryptographic module, the form in which they are entered, and the method used for entry[14].

FR: 6.59 The CKMS design **shall** specify how the integrity and confidentiality (if necessary) of the entered keys and metadata are protected and verified upon entry.

6.4.20 Output a Key and Associated Metadata from a Cryptographic Module

The key output function outputs one or more keys and associated metadata from a cryptographic module for external use or storage. Output may be for archive, backup, or normal, operational purposes. A module that serves as a key generation facility may output keys for subsequent distribution. Keys and metadata may be output in plaintext form, in encrypted form, as key splits, in integrity-protected form, or any combination thereof.

FR: 6.60 The CKMS design **shall** specify how, and the circumstances under which, keys and metadata are output from a cryptographic module and the form in which they are output.

FR:6.61 The CKMS design **shall** specify how the confidentiality and integrity of the output keys and metadata are protected while outside of a cryptographic module.

FR: 6.62 If a private key, symmetric key, or confidential metadata is output from the cryptographic module in plaintext form, the CKMS design **shall** specify if and how the calling entity is authenticated before the key and metadata are provided.

6.4.21 Validate Public Key Domain Parameters

This function performs certain validity checks on the public domain parameters of some public key algorithms. Passing these tests provides assurance that the domain parameters are arithmetically correct (see [SP 800-89] and [SP 800-56A]).

---

[14] For example, by keyboard entry, key loader, or via automated protocols.

FR: 6.63 The CKMS design **shall** specify how, where, and the circumstances under which, public key domain parameters are validated.

6.4.22 Validate Public Key

This function performs certain validity checks on a public key to provide some assurance that it is arithmetically correct. These tests typically depend on the public key algorithm for which the key is intended, but do not depend on knowledge of the private key (see [SP 800-89], [SP 800-56A], and [SP 800-56B]). Note that Sections 6.4.22, 6.4.23, and 6.4.28 are related to providing an overall trust scenario for the validation of these keys.

FR: 6.64 The CKMS design **shall** specify how, where, and the circumstances under which, public keys are validated.

6.4.23 Validate Public Key Certification Path

This function validates the certification path (also known as a certificate chain), from the trust anchor of the relying entity to a public key in which the relying entity needs to establish trust (i.e., the public key of the other entity in a transaction). The validation of the certification path provides assurance that the subject identity that is given in the certificate is, in fact, the identity of the owner of the static public key and the holder of the corresponding static private key (assuming that proof of private key possession was verified by the certificate authority or some other entity trusted by the relying entity).

FR: 6.65 The CKMS design **shall** specify how, where, and the circumstances under which, a key certification path are validated.

6.4.24 Validate Symmetric Key

This function performs tests on the symmetric key and its metadata. For example, tests may include checking for the proper length and format of the key. This command may also verify any error detection/correction codes or integrity checks placed upon the key and/or its metadata.

FR: 6.66 The CKMS design **shall** specify how, where, and the circumstances under which, symmetric keys and/or metadata are validated.

6.4.25 Validate Private Key (or Key Pair)

This function performs certain tests on a private key to provide assurance that it meets its specifications. The test can only be performed by the private-key owner or by a trusted third-party acting on behalf of the private-key owner. This test may also involve a pair-wise consistency test that verifies that the private key performs a complementary function to the public key. For example, in the case of an RSA key pair, applying the private key to a given input block, followed by applying the public key to the result should always yield the given input block (see Section 6.4.1 of [SP 800-56B] for more information).

FR:6.67 The CKMS design **shall** specify how, where and the circumstances under which, private keys or key pairs and/or metadata are validated.

6.4.26 Validate the Possession of a Private Key

This function is used by an entity that receives a public key and wishes to obtain assurance that the claimed owner of the public key has possession of the corresponding private key, and is therefore the owner of the key pair. The key-pair owner is typically required to use the private key in a cryptographic transaction in which another entity uses the public key in an attempt to verify the possession. For example, the owner may sign data (e.g., the public key and other information) using the private key before sending it to the receiver. The receiver uses the received public key to validate the signature on the received data (see [SP 800-56A], [SP 800-56B], and [SP 800-89]). This function may also contain the capability for a private-key owner to validate the possession of the owner's own private key.

FR: 6.68 The CKMS design **shall** specify how, where, and the circumstances under which, possession of private keys and their metadata are validated.

6.4.27 Perform a Cryptographic Function using the Key

The main usage functions are the actual functions that provide the cryptographic protection to data. These functions may include signature generation, signature verification, encryption, decryption, key wrapping, key unwrapping, MAC generation, and MAC verification. They should be performed within a cryptographic module.

FR: 6.69 The CKMS design **shall** specify all cryptographic functions that are supported and where they are performed in the CKMS (e.g., CA, host, or end user system).

6.4.28 Manage the Trust Anchor Store

A CKMS may require that certain entities have one or more trusted public keys. These public keys are also referred to as trust anchors. A trust anchor is used to establish trust in other public keys that are not otherwise trusted. The trust in these otherwise un-trusted public keys is established by verifying all signatures in a chain of public key certificates (termed "certification path" in Section 6.4.23), starting with a trust anchor that is trusted by the relying entity. Thus, the integrity of trust anchors is critical to the security of the CKMS. The CKMS typically supports trust anchor management functions, such as adding, deleting and storing trust anchors. Trust anchor formats are specified in [RFC 5914]. The Secure Trust Anchor Management Protocol (TAMP) is defined in [RFC 5934].

FR: 6.70 The CKMS design **shall** specify all trust anchor management functions that are supported (see [RFC 6024]).

FR: 6.71 The CKMS design **shall** specify how the trust anchors are securely distributed so that the relying parties can perform source authentication and integrity verification on those trust anchors.

FR: 6.72 The CKMS design **shall** specify how the trust anchors are managed in relying-entity systems to ensure that only authorized additions, modifications, and deletions are made to the relying-entity system's trust anchor store.

6.5 Cryptographic Key and/or Metadata Security: In Storage

When cryptographic keys are submitted for storage, they are typically submitted with their metadata. The metadata may include an owner identifier or user access control list. If any of the metadata is incorrect, then the false information will be perpetuated by the CKMS system. Therefore, a CKMS storage system should verify the authorization of the submitting entity and the integrity of the submitted data before any data is stored[15].

When cryptographic keys are stored, they require protection. Symmetric keys and private keys require confidentiality protection and access control. All keys require integrity protection. For confidentiality protection, cryptography, computer security, and/or physical security can be employed. If symmetric key cryptography is used for key confidentiality, then there often exists a symmetric key wrapping key that is used to encrypt and decrypt the stored keys and confidential metadata. At the top level in the key encrypting key hierarchy, there typically is a key that must be physically protected.

If asymmetric key cryptography is used for key confidentiality, then a public key could be used to encrypt stored keys. The corresponding private key that is used to decrypt the keys must be protected in some manner, e.g., using physical security and key splitting (see Section 6.7.5), that usually does not involve encryption.

All stored keys require integrity protection because a garbled key will not correctly perform its intended function and may compromise another key under some circumstances. Physical security can provide integrity protection for keys, but additional methods are frequently used. An error detection code can detect an unintentional garble in a key, and an error correction code can correct certain garbles. However, if a key could be intentionally garbled, then a cryptographic integrity check like a MAC or digital signature should be implemented for error detection. If an uncorrectable garble is detected, the garbled key should not be used. When public keys are contained within a certificate, they are provided integrity protection by means of the digital signature on the certificate. If public keys are stored outside of their certificate, then their integrity needs to be protected by some other means.

A CKMS should only allow authorized users to have access to stored keys. Thus, a CKMS should have some type of access control system (ACS) (see Section 6.7.1). The ACS may be as simple as requiring a password or cryptographic key from the authorized user of the key, and/or it may make use of biometric authentication techniques.

---

[15] It is also a good practice to verify the integrity of keys and metadata immediately upon access and before operational use.

A key may be garbled, lost, or destroyed to the extent that it cannot be reconstructed by error correction codes. If the key is a symmetric key or a private key, this could result in the loss of the data protected by the key. A CKMS should employ methods for backing-up, archiving, and recovering keys as necessary to provide for the recovery of valuable data. For example, Appendix B of [SP 800-57-part1] provides guidance on recovery procedures for various key types.

A garble in key metadata could result in the misuse of the key or the denial of service. Therefore, metadata may also require backup, archiving, and recovery.

FR: 6.73 The CKMS design **shall** specify the methods used to authenticate the identity and verify the authorization of the entity submitting keys and/or metadata for storage.

FR: 6.74 The CKMS design **shall** specify the methods used to verify the integrity of keys and/or metadata submitted for storage.

FR: 6.75 The CKMS design **shall** specify the methods used to protect the confidentiality of symmetric and private stored keys and metadata.

FR: 6.76 If a key wrapping key (or key pair) is used to protect stored keys, then the CKMS design **shall** specify the methods used to protect the key wrapping key (or key pair) and control its use.

FR: 6.77 The CKMS design **shall** specify the methods used to protect the integrity of stored keys and metadata.

FR: 6.78 The CKMS design **shall** specify how access to stored keys is controlled.

FR: 6.79 The CKMS design **shall** specify the techniques used for correcting or recovering all stored keys.

6.6 Cryptographic Key and Metadata Security: During Key Establishment

Keys and metadata can be established between entities wishing to communicate using key transport or key agreement methods. These methods are typically used to establish keys over electronic communications networks, but they could also be used to provide extra security (beyond physical protection) when keys are manually distributed. When keys are transported, one entity generates the key to be shared, and the key and possibly its metadata are distributed to the other entity. When keys are agreed upon, both entities contribute information that is used to derive a shared key. Metadata may be transported under the protection of the shared key. [SP 800-56A] and [SP 800-56B] specify cryptographic schemes for key establishment.

6.6.1 Key Transport

When cryptographic keys and metadata are transported (distributed) from one entity (the sender) to another (the intended receiver), they should be protected. Symmetric keys and private keys require confidentiality protection and access control. For confidentiality protection, either physical security or cryptography is used. A manually distributed key could be physically protected by a trusted courier, or a physically protected channel could be used. Very often, the keys are sent electronically over networks that are susceptible to data eavesdropping and modification. If cryptography is used to protect the confidentiality of symmetric and private keys during transport, then a key establishment technique involving either a symmetric key wrapping key or one or more asymmetric transport key pairs is used. These wrapping and transport keys also should be protected by the end entities involved in the transport.

All transported keys require integrity protection because a garbled key will not correctly perform its intended function, and attacker-controlled key garbles could result in spoofing or cryptanalytic attacks. Thus, detecting garbled keys prior to their use improves the security and reliability of the system. Physical security can provide integrity protection for keys, but often other methods are used, due to the lack of physical protection of electronic data on typical networks. An error detection code can detect an unintentional garble to a key, and an error correction code can correct certain garbles. However, if a key could be intentionally garbled, then a cryptographic integrity check, like a MAC or digital signature, should be used for error detection. If an uncorrectable garble is detected, a new or corrected key should be established before use. When public keys are contained within a certificate, they are provided integrity protection by the digital signature on the certificate.

The receiver of a transported key desires assurance that the key came from the expected authorized key sender. When transported using automated methods, this assurance is typically provided by the use of a cryptographic mechanism that authenticates the identity of the sender to the receiver. When a key is transported manually, this assurance may be provided by the authentication of the trusted courier who transports the key.

FR: 6.80 The CKMS design **shall** specify the methods used to protect the confidentiality of symmetric and private keys during their transport.

FR: 6.81 The CKMS design **shall** specify the methods used to protect the integrity of transported keys and how the keys can be reconstructed or replaced after detecting errors.

FR: 6.82 The CKMS design **shall** specify how the identity of the key sender is authenticated to the receiver of transported keying material.

6.6.2 Key Agreement

Two entities, working together, can create and agree on a cryptographic key without the key being transported from one to the other. Each entity supplies some information that is used to derive a common key, but when secure key agreement schemes are used, an eavesdropper obtaining this information is not able to determine the agreed-upon key. Cryptographic algorithms employing key agreement keys are used by each entity.

Each entity participating in a key agreement process typically needs assurance as to the identity of the other entity. This assurance is often provided by the key agreement protocol.

FR: 6.83 The CKMS design **shall** specify each key agreement scheme supported by the CKMS.

FR: 6.84 The CKMS design **shall** specify how each entity participating in a key agreement is authenticated.

6.6.3 Key Confirmation

When keys are established between two entities, each entity may wish to have confirmation that the other entity did, in fact, establish the correct key. Key confirmation schemes are used to provide this capability. [SP 800-56A] and [SP 800-56B] specify key confirmation schemes for use in Federal CKMS. Other methods may also be appropriate.

FR: 6.85 The CKMS design **shall** specify each key confirmation method used to confirm that the correct key was established with the other entity.

FR: 6.86 The CKMS design **shall** specify the circumstances under which each key confirmation is performed.

6.6.4 Key Establishment Protocols

Several automated protocols have been developed for the provision of cryptographic keys for both storage and transmission. Often, these protocols are designed for a particular application or set of applications. Some well-known key establishment protocols include:

    a) Internet Key Exchange (IKE)
    b) Transport Layer Security (TLS)
    c) Secure/Multipart Internet Mail Extensions (S/MIME)
    d) Kerberos
    e) Over-The-Air-Rekeying (OTAR) Key Management Messages
    f) Domain Name System Security Extensions (DNSSEC)
    g) Secure Shell (SSH)

A high-level overview of items a) through f) can be found in [SP 800-57-part3], along with guidance as to which cryptographic options are recommended for U.S. Government use. For Secure Shell information, see [RFC 4251].

FR: 6.87 The CKMS design **shall** specify all the protocols that are employed by the CKMS for key establishment and storage purposes.

6.7 Restricting Access to Key and Metadata Management Functions

This section describes how access to the key and metadata management functions may be controlled. The requesting entity may be authenticated, and human exposure to keys and other sensitive metadata may be prevented or severely restricted.

6.7.1   The Access Control System (ACS)

The security of a CKMS depends on the proper sequence and execution of the key and metadata management functions described in Section 6.4. The execution of these functions may be driven by time, an event, an entity's request, or some combination of these options. An access control system is necessary to assure that key and metadata management functions are only performed in response to requests (calls) by authorized entities[16] and that all applicable constraints are met[17]. For example, the recover key function (see Section 6.4.17) may be restricted to the cryptographic officer role, and input parameters may be verified to be within specified bounds and have specified formats.

The Access Control System works in conjunction with cryptographic modules to control access to sensitive keys and metadata. An Access Control System (ACS) protects keys by ensuring that only authorized entities are permitted to execute key and metadata management functions. In addition, administrative access is typically logged and audited for personal accountability. An ACS could be very simple; for example, any user submitting an appropriate identifier and password might be authorized to perform any key management function with any key, or the ACS may be much more complex.

Figure 10 illustrates the relationships between the calling entity, the Access Control System, protected memory, and the cryptographic module. These devices may be collocated, or they may be connected by a secure channel as shown in the figure. A calling entity makes CKMS function calls that are serviced by the ACS. The ACS makes use of protected memory and a cryptographic module to authenticate the calling entity. If the authentication is successful, and the entity is properly authorized, then the function is performed by making cryptographic service requests to the cryptographic module. Finally, the response is then passed back to the calling entity.

Additional details of a sample key management function operation are shown in Figure 11. A function call consisting of the calling entity's identifier, the calling entity's authenticator, the function name, and the key identifier is presented to the ACS. The entity is first authenticated. Then the entity's authorization to exercise the command is verified by checking that the entity's ID is in the access control list (located in the key metadata) for the key and the function. If the ACS determines that the function should not be permitted, then it returns a function-denied indicator. If, however, the function is permitted for the authenticated entity using the key and metadata, then the ACS notifies the function logic to perform the requested operation. The function logic may call upon the cryptographic module to encrypt, decrypt, sign, verify or compute a MAC, as necessary. Finally, the response to the function call is provided to the calling entity.

---

[16] The authorization of an entity is determined after the identity (or role) of the entity is authenticated. The identity (or role) is verified as approved to execute the function.

[17] Constraints are limitations that are placed upon the input to and use of the function to help ensure correct and secure operation.

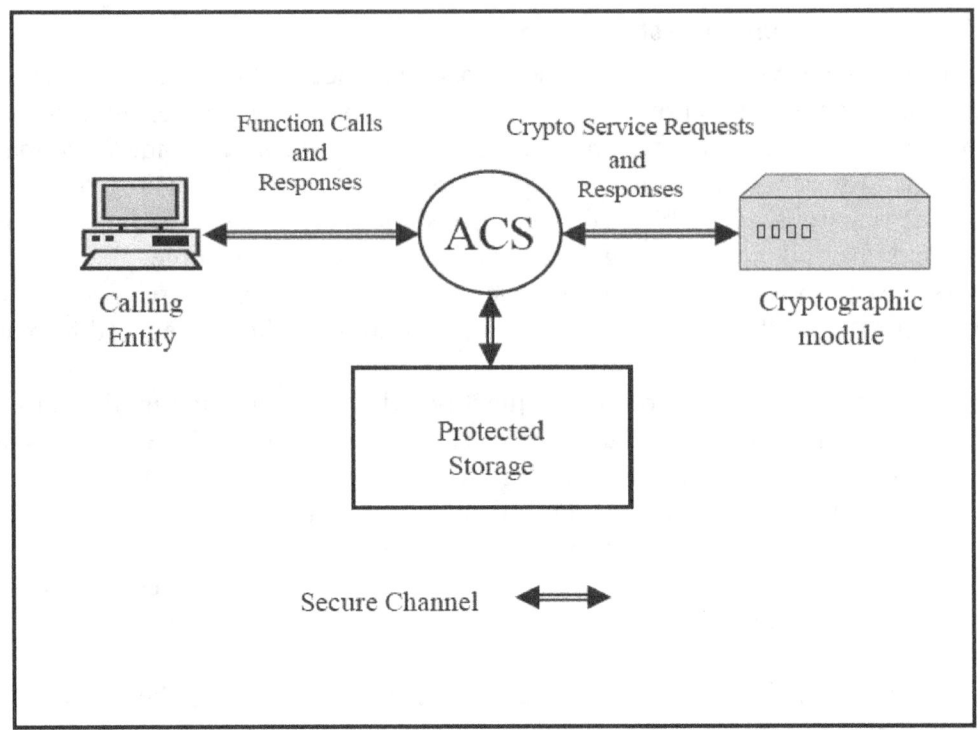

Figure 10: Management Function Access Control

The ACS makes the decision to perform the requested function or not. This decision is primarily based on the authenticated identity (or role) of the calling entity, the authorizations of the entity, the security policies of the CKMS, the function, the key, and its metadata. The metadata of a key may play a critical role in determining the controls that are to be enforced. For example, an organization may decide that multiple users will be permitted to use a shared key to encrypt and decrypt a particular file, while another file can be decrypted only by a single user. The CKMS policies should support and enforce the information management policies of the managing organization. Therefore, it is highly desirable that a CKMS access control system be flexible enough to accommodate the requirements of the CKMS Security Policy.

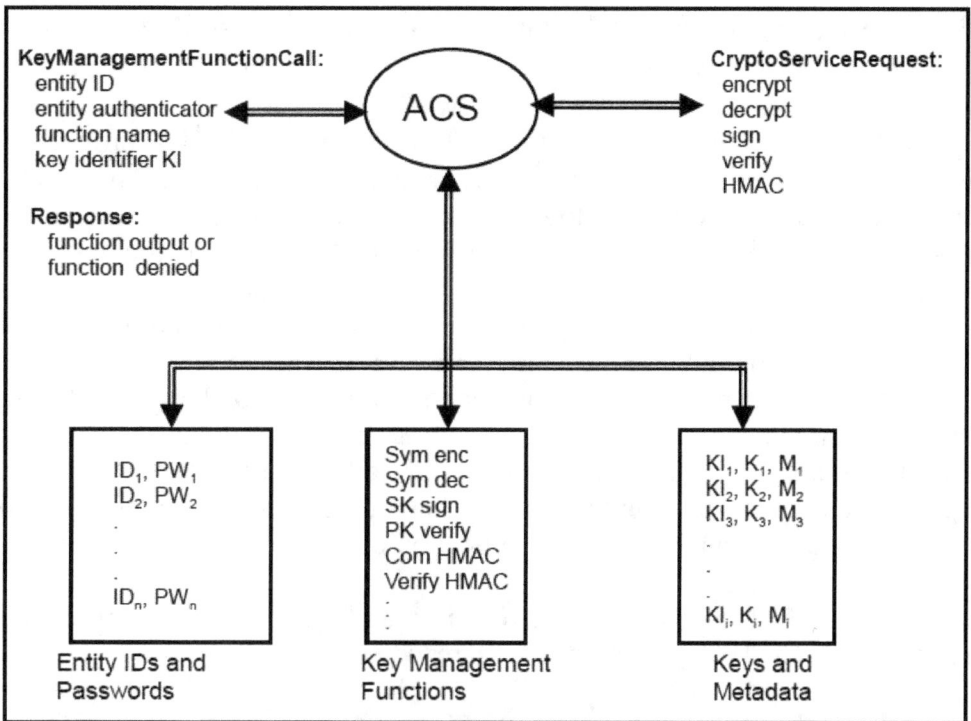

Figure 11: Sample Key Management Function Control Logic

FR: 6.88 The CKMS design **shall** specify the topology of the CKMS by indicating the locations of the entities, the ACS, the function logic, and the connections between them.

FR: 6.89 The CKMS design **shall** specify the constraints on the key management functions that are implemented to assure proper operation.

FR: 6.90 The CKMS design **shall** specify how access to the key management functions is restricted to authorized entities.

FR: 6.91 The CKMS design **shall** specify the ACS and its policy for controlling access to key management functions.

FR: 6.92 The CKMS design **shall** specify at a minimum:
    a) The granularity of the entities (e.g., person, device, organization),
    b) If and how entities are identified,
    c) If and how entities are authenticated,
    d) If and how the entity authorizations are verified, and
    e) The access control on each key management function.

FR: 6.93 The CKMS design **shall** specify the capabilities of its ACS to accommodate, implement, and enforce the CKMS Security Policy.

6.7.2   Restricting Cryptographic-Module Entry and Output of Plaintext Keys

A well-designed CKMS will minimize the access of humans to plaintext keys. The primary need for keys to be in plaintext is when they are performing cryptographic functions within a cryptographic module. These modules usually provide physical protection to the plaintext keys so that they will not be exposed. The module may generate the keys and perform cryptographic functions on behalf of humans, and the humans need never see a plaintext symmetric or private key. This feature makes a CKMS using such modules more transparent and more secure. For example, a private key transport key could be generated within the module and never be allowed to leave the module. Keys that are output from the module could be transported (in encrypted form) using a key transport scheme. A symmetric encryption/decryption key could be encrypted and transported using the public key of the receiving entity. A key may be securely stored outside of the module when encrypted under a public key storage key or symmetric key wrapping key. Sometimes, plaintext key output is permitted to support legacy systems. In such cases, multi-party control, discussed in Section 6.7.4 below, should be considered.

Requirements for the entry and output of keys into and from a cryptographic module are specified in Section 6.4.19 and Section 6.4.20, respectively.

FR: 6.94 The CKMS design **shall** specify the circumstances under which plaintext secret or plaintext private keys are entered into or output from a cryptographic module.

FR: 6.95 If plaintext secret or plaintext private keys are entered into or output from any cryptographic module, then the CKMS design **shall** specify how the plaintext keys are protected and controlled outside of the cryptographic module.

FR: 6.96 If plaintext secret or plaintext private keys are entered into or output from any cryptographic module, then the CKMS design **shall** specify how such actions are audited.

6.7.3   Controlling Human Input

If a key management function requires the human input of keys or sensitive metadata, then there is a dependence on the human for the accuracy and perhaps the security of the input. In addition, there could be a dependency on the human to enter the input at the proper time or when the proper event occurs. In this case, the issue arises as to what action the system should take if the human input is not provided. If such functions can be performed automatically by the CKMS when they are necessary, the system becomes more transparent to the user and possibly more secure.

FR: 6.97 For each key and metadata management function, the CKMS design **shall** specify all human input parameters, their formats, and the actions to be taken by the CKMS if they are not provided.

6.7.4   Multiparty Control

Certain key management functions could require multiple cooperating entities to perform the function. This multiparty control could be enforced by requiring k of n entities to authenticate to and be authorized by the function's access control system before the function is performed. Multiparty controls should be used for highly sensitive functions. For example, a highly sensitive function should require that two or more individuals be logged on and authenticated to perform the function.

FR: 6.98 The CKMS design **shall** specify all functions that require multiparty control, specifying k and n for each function.

FR: 6.99 For each multiparty function, the CKMS design **shall** cite or specify any known rationale (logic, mathematics) as to why any k of the n entities can enable the desired function, but $k-1$ of the entities cannot.

6.7.5   Key Splitting

Key splitting is a technique for multiparty control. When a highly sensitive key is required, n key splits are generated so that any k of the key splits can be used to form the key, but any $k-1$ key splits provide no knowledge of the key. Each of the n key splits is then assigned to one of n trusted entities so that the key cannot be formed unless k of the entities agree to take part. If any $k-1$ of the entities had their key splits compromised, the key could still not be reconstructed by an attacker having all the $k-1$ key splits. Thus, the security of the key is distributed. Split knowledge procedures have been used to establish root or master keys that provide protection to many other keys and whose compromise would result in a major disaster. These key splits (rather than the plaintext key resulting from combining the key splits) are often entered into, or output from, the CKMS in plaintext form for backup purposes.

FR:6.100 The CKMS design **shall** specify all keys that are managed using key splitting techniques and **shall** specify n and k for each technique.

FR:6.101 For each (k, n) key splitting technique used, the CKMS design **shall** specify how key splitting is done, and any known rationale (logic, mathematics) as to why any k of the n key splits can form the key, but $k-1$ of the key splits provide no information about the key.

6.8 Compromise Recovery

In an ideal situation, the CKMS would protect all keys and sensitive metadata so that they are never compromised or modified by unauthorized parties. However, since it is difficult or even impossible to design a perfect CKMS that prevents all potential security problems, a CKMS should be designed to detect compromises and unauthorized modifications, to mitigate their undesirable effects, to alert the appropriate parties of compromises, and to recover (or help recover) to a secure state once a compromise or unauthorized modification is discovered. This section addresses how the recovery from a compromise should occur.

When a CKMS compromise is detected
   a) The compromise should be evaluated to determine its cause and scope,

b) Compromise-mitigation measures should be instituted to minimize key and/or metadata exposure,

c) Appropriate corrective measures should be instituted to prevent the reoccurrence of the compromise, and

d) The CKMS should be returned to a secure operating state.

6.8.1 Key Compromise

Depending on the key type and key usage, the compromise of a key could result in

a) Loss of confidentiality,

b) Loss of integrity,

c) Loss of authentication,

d) Loss of non-repudiation, or

e) Some combination of these losses.

Note that the loss of a security service provided to a key is likely to result in a loss of the same and potentially other security services for data protected by the key. For example, a loss of the integrity for a public key transport key could impact the confidentiality of the data encryption key protected by the public key and that, in turn, could compromise the confidentiality of the data protected by the data encryption key. (More specifically, if a public RSA key is changed to have the value 1 modulo n, then any data encrypted by that altered key would be compromised.)

A key compromise could be undetected, detected or suspected. A CKMS should limit the exposure of undetected key compromises by establishing a cryptoperiod or usage limit for each key that it uses[18]. At the end of each cryptoperiod, a new key could be established to replace the old key. When a new key is established and activated to protect new data, the old key should no longer be used to protect the new data. Thus, unless the compromise recurs with the new key, the new data will be protected. Of course, the old data that was protected with the old key could have been compromised, but the extent of the compromise is limited, as long as the old key was not used to protect the new key (e.g., the old key was not used to protect the new key during key transport). If a key compromise is detected, then the compromised key and other keys whose security depends upon the security of the compromised key should be replaced as soon as possible. Since the compromise of a key may result in the compromise of many other keys that it protects, it is important to design a CKMS to minimize the impact of key compromise. [SP 800-57-part1] provides guidance as to appropriate cryptoperiods for various key types.

If a symmetric key wrapping key, a private key transport key, or a private key agreement key is compromised, then transported or agreed-upon keys might be compromised as well. If the compromise is undetected, the compromise of additional keys might continue indefinitely. Some protocols are designed to prevent or mitigate such attacks. However it

---

[18] The usage of keys may be limited based on a criterion such as the amount of data processed using the key or the number of times the algorithm was initialized using the key.

is generally considered a good idea to keep the cryptoperiods of the symmetric key wrapping, key transport, and key agreement keys to the minimum practical period of time.

If a key derivation key or master key is compromised, then any key derived from the key derivation or master key could also be compromised. Therefore, key derivation and master keys should also be changed on a periodic basis.

FR: 6.102 The CKMS design **shall** specify the range of acceptable cryptoperiods or usage limits of each type of key used by the system.

FR:6.103 For each key, a CKMS design **shall** specify the other key types that depend on the key for their security and how those dependent keys are to be replaced in the event of a compromise of the initial key.

FR: 6.104 The CKMS design **shall** specify the means by which other compromised keys can be identified when a key is compromised. For example, when a key derivation key is compromised, how are the derived keys determined?

6.8.2   Metadata Compromise

Depending on the metadata element and how it is used, the compromise of a metadata element could result in the compromise of a key or the data protected by a key. For example, a metadata element of a symmetric encryption/decryption key could be a list of identities corresponding to the legitimate users of the key. The Access Control System verifies the authenticated identity of the user against the metadata element to determine whether the user is permitted to exercise the decrypt function and thus obtain plaintext data. If the metadata element could be modified to add an unauthorized user to the list of authorized users, then the encrypted data could be compromised. If different keys have common metadata elements, then the compromise of one metadata element could compromise the data protected by each of the keys. Metadata elements that are sensitive to unauthorized modification should be cryptographically bound to their associated keys so that the integrity of the metadata can be easily verified.

FR: 6.105 For each key type employed, the CKMS design **shall** specify which metadata elements are sensitive to compromise (confidentiality, integrity, or source).

FR: 6.106 The CKMS design **shall** specify the potential security consequences, given the compromise (confidentiality, integrity or source) of each sensitive metadata element of a key.

FR: 6.107 The CKMS design **shall** specify how each sensitive metadata element compromise can be remedied.

6.8.3 Key and Metadata Revocation

Keys are revoked for a number of reasons, including key compromise and the termination of an employee or the employee's role within an organization. A CKMS should have the ability to rapidly replace keys (both asymmetric and symmetric) and the ability to notify the relying parties (those who make use of the key) of compromise/revocation.

Compromised Key Lists (CKLs), Certificate Revocation Lists (CRLs) (see [RFC 5280]), White Lists, Query White Lists, and the Online Certificate Status Protocol (OCSP) (see [RFC 6960]) are examples of mechanisms in use for the promulgation of key revocation information to the relying entities. Each mechanism has its benefits and drawbacks. For example, CRLs and CKLs have problems with growing very large and becoming out of date (i.e., stale). Growth adversely impacts communication, computing, and storage requirements. The growth problems for the end entity can be mitigated by partitioning the revocation information into smaller chunks, each chunk handling fewer keys. Staleness cannot be fully eliminated, but can be mitigated by issuing lists more frequently. Note that in some instances, more than one revocation mechanism can be used to meet the security requirements and limitations of the relying parties.

Key revocation mechanisms should consider:
a) Relying entity requirements for the timeliness of revocation information,
b) Relying entity computing and communication limitations, and
c) Infrastructure cost considerations.

FR:6.108 A CKMS design **shall** specify the key revocation mechanism(s) and associated relying entity notification mechanism(s) used or available for use.

6.8.4 Cryptographic Module Compromise

Since a cryptographic module contains plaintext keys at some point in time, the compromise of the module has the potential to compromise the symmetric and private keys contained within the module (see Section 8.4). This could lead to the loss of confidentiality, the loss of integrity, or the loss of the ability to authenticate, as described in Section 6.8.1 above. Cryptographic modules can be compromised either physically (i.e., obtaining direct access to the keys within the module) or by non-invasive methods so that knowledge of the keys within the module is obtained by some external action. To provide physical protection, modules should operate in a space where unauthorized access is not permitted or where unauthorized access is quickly detected before a serious compromise occurs. Some modules provide this protection at their cryptographic boundary, but larger boundaries may also be involved. See [FIPS 140] for more information on the physical protection of a cryptographic module's contents. If access to the contents of a cryptographic module is permitted, then an access control system may be required to ensure that only authorized parties succeed.

Following an actual or suspected cryptographic module compromise, a secure state of the module should be re-established. The actions required to return to this state are collectively called recovery. Recovery sometimes requires the replacement of internal hardware and/or software of the module. The module should be returned to a secure state before the module is returned to normal operation. Following repair or replacement, a module must be tested for its operational capability, as well as its security status.

To provide protection against non-invasive attacks on a cryptographic module, either the use of the module should be restricted to only trusted users, or the module should be designed to prevent this specific type of attack. In the first case, there is always the threat that a module will be lost or stolen or that a trusted user will become dishonest. In the second case, it can become very costly to protect against every possible type of non-invasive attack. An attacker might determine information about a cryptographic key used by the module by examining the detailed power consumption patterns of the module during the cryptographic processing. Other potential non-invasive attacks are based on carefully analyzing the amount of time certain cryptographic functions take to execute, or the emanations given off by the module during its normal operation.

FR:6.109 The CKMS design **shall** specify how physical and logical access to the cryptographic module contents is restricted to authorized entities.

FR:6.110 The CKMS design **shall** specify the approach to be used to recover from a cryptographic module compromise.

FR:6.111 The CKMS design **shall** describe what non-invasive attacks are mitigated by the cryptographic modules used by the system and provide a description of how the mitigation is performed.

FR:6.112 The CKMS design **shall** identify any cryptographic modules that are vulnerable to non-invasive attacks.

FR:6.113 The CKMS design **shall** provide the rationale for accepting the vulnerabilities caused by possible non-invasive attacks.

6.8.5   Computer System Compromise Recovery

The unauthorized modification of CKMS software or major portions of a computer operating system can be detected using tools that run on a separate secure platform and monitor any modification to a file, changes to the hash value of a file's contents, or changes to a file's attributes (e.g., who the owner is, or who is on the ACL) (see Section 8.2.4). Alternatively, a layered system of protections is often built into a CKMS. When protective mechanisms are built into the system, they need to be protected from the same threats as the system itself. When critical files undergo unauthorized modifications that are detected by the monitoring utility or indicated in the event log, these files should be replaced using known valid and secure files located in secure storage.

If pervasive, unauthorized changes to software are made, the software should be recovered as described in Section 10.5.

FR:6.114 The CKMS design **shall** specify the mechanisms used to detect unauthorized modifications to the CKMS system hardware, software and data.

FR:6.115 The CKMS design **shall** specify how the CKMS recovers from unauthorized modifications to the CKMS system hardware, software and data.

6.8.6   Network Security Controls and Compromise Recovery

The compromise of network security controls that provide protection to the CKMS could result in the compromise of the CKMS itself. The scope of network security controls includes boundary devices, such as a firewall, a VPN, an intrusion detection system, and an intrusion protection system. The scope of network security controls excludes cryptographic functions, cryptographic protocols, and cryptographic services, except when used for the operation of the aforementioned network security control devices.

The following are some of the examples of compromises of network security controls:

a) The physical compromise of a network security control device,
b) A compromise of one or more cryptographic keys used by a network security control device,
c) A compromise of one or more keys used to administer the network security control device,
d) A change in the network architecture resulting in a compromise (e.g., someone connecting a VPN-connected workstation to an unsecure network and the VPN workstation being used to attack the Intranet),
e) A compromise of a privileged user password (e.g., a system administrator's password),
f) A compromise of a platform operating system,
g) A compromise of a network security application (e.g., a firewall, IDS, etc.), and
h) A compromise due to a new attack on a protocol.

If physical security is compromised, the device should be replaced with a new device and physical security controls should be reviewed, repaired, and enhanced, as appropriate.

If device or administration keys are compromised, the keys should be replaced. An assessment should be conducted to determine the cause of the compromise, the extent of the damage, and corrective actions should be taken. In the unlikely event of the security strength of the key being an issue, the key sizes may need to be increased and/or more secure cryptographic algorithms may need to be used.

If the network architecture assumptions are violated, the cause of the violation should be reviewed, and appropriate actions should be taken.

Compromised network devices should be restored to a secure state before normal operation is continued.

If passwords are compromised, the passwords should be replaced. The users may require further training in selecting the password, in understanding password entropy, in changing passwords frequently, and in maintaining the confidentiality of written-down passwords. An examination should also be made of the authentication protocols to determine if password sniffing, online dictionary attacks or offline dictionary attacks are feasible.

If the platform operating system is compromised, one or more of the following actions should be considered and appropriate corrective measures taken:
   a) Make sure that all the latest operating system security patches are installed,
   b) Ask the operating system vendor if there is a patch for the compromise, or
   c) Determine if a device configuration change or the blocking of some protocols will prevent future attacks of the same nature as the one that caused the compromise.

If the network security application is compromised, one or more of the following actions should be considered, and appropriate corrective measures should be taken:
   a) Make sure that all the latest network security patches are installed,
   b) Ask the application vendor if there is a patch for the compromise, or
   c) Determine if a device change, an application configuration change, or the blocking of certain protocols will prevent future attacks that allowed or caused the compromise.

If the compromise is due to an inadequate network security protocol, one or more of the following actions should be considered, and appropriate corrective measures should be taken:
   a) Ask the network security application vendor if there is a patch for the compromise, or
   b) Determine if a device configuration change or the blocking of certain protocols will prevent future attacks of the same nature as the one that caused the compromise.

In all of these situations, the incident should be fully investigated to determine what other systems and keys may have been compromised due to a compromise of network security controls. This damage assessment could lead to additional compromise declarations and additional compromise recovery procedures.

FR:6.116 The CKMS design **shall** specify how to recover from the compromise of the network security control used by the system. Specifically,
   a) The CKMS design **shall** specify the compromise scenarios considered for each network security control device,
   b) The CKMS design **shall** specify which of the mitigation techniques specified in this section are to be employed for each envisioned compromise scenario, and
   c) The CKMS design **shall** specify any additional or alternative mitigation techniques that are to be employed.

6.8.7 Personnel Security Compromise Recovery

The humans who are responsible for the correct and secure operation of a CKMS often have the capability to compromise its security. However, a CKMS can be designed with its own capabilities to minimize the likelihood of human compromises, detect the compromises, minimize the negative consequences of the compromises, and efficiently recover from the compromises.

A CKMS should be designed to:
   a) Minimize the ability of humans to cause security failures,
   b) Minimize the ability of humans to hide their actions that caused security failures,
   c) Help determine who or what caused the security failure (for example by maintaining audit records), and
   d) Mitigate the negative consequences of the failure.

Any detected security failure should result in the initiation of recovery procedures based upon the Information Security Policy and the CKMS capabilities.

Typical responses include:
   e) The complete shut-down of the system,
   f) The activation of a backup facility and system with new keys,
   g) The notification of current and potential users of the possible security failure, or
   h) The flagging of the keys that were compromised.

In addition to the above responses, failures involving personnel compromise could vary from administrative reprimands, to removal from the role or position and legal action involving civil or criminal courts.

FR:6.117 The CKMS design **shall** specify any personnel compromise detection features that are provided for each supported role.

FR: 6.118 The CKMS design **shall** specify any personnel compromise minimization features that are provided for each supported role.

FR:6.119 The CKMS design **shall** specify the CKMS compromise recovery capabilities that are provided for each supported role.

6.8.8 Physical Security Compromise Recovery

The physical security of a cryptographic module is discussed in Section 6.8.4, and the general compromise of keys and metadata is discussed in Section 6.8.1 and Section 6.8.2, respectively. However a physical security breach of a CKMS could involve compromises other than the compromise of keys or cryptographic modules. If security-related logic resides outside of the CKMS cryptographic modules, then the integrity of that logic also should be protected. Typically, techniques similar to those used by the cryptographic module are employed. Physical protection can be provided that prevents the potential attacker from gaining physical access to the components and devices. Alternatively, detection mechanisms could be used to detect an unauthorized access and then take some defensive action. For example, a detected unauthorized access could sound an alarm or send an alert to the security officer. Often, a combination of prevention and detection measures is used.

Once security is breached, the integrity of the entire breached area should be suspect. If the CKMS detects a breach, it should inform the appropriate entity about the breach, as specified in the CKMS Security Policy, so that mitigation actions can be taken. In addition, it might not be sufficient to replace all sensitive data within the breached area, because the attacker could have modified or added to the logic within the area so that the new keys and sensitive information could also be compromised in the future.

FR:6.120 The CKMS design **shall** specify how all CKMS components and devices are protected from unauthorized physical access.

FR:6.121 The CKMS design **shall** specify how the CKMS detects unauthorized physical access.

FR:6.122 The CKMS design **shall** specify how the CKMS recovers from unauthorized physical access to components and devices other than cryptographic modules.

FR:6.123 The CKMS design **shall** specify the entities that are automatically notified if a physical security breach of any CKMS component or device is detected by the CKMS.

FR:6.124 The CKMS design **shall** specify how breached areas can be re-established to a secure state.

7. Interoperability and Transitioning

Interoperability is the ability of diverse systems to communicate and work together (i.e., interoperate)[19]. A CKMS may interoperate with an application or a peer CKMS. Interoperability can only be achieved by having a detailed specification of the interfaces to systems with which the CKMS intends to interoperate. This inherently includes the following:

a) Common interfaces and protocols, i.e., the syntax and semantics of interfaces that invoke functions and services from one CKMS entity to another CKMS entity are the same for interoperating systems,

b) Formats for keys, metadata, and other exchanged data are the same or can be translated by interoperable systems, and

c) Data exchange mechanisms, including security mechanisms, are the same or are equivalent between interoperable systems.

---

[19] See http://en.wikipedia.org/wiki/interoperability for more information on the power and uses of interoperability.

Current cryptographic algorithms should be implemented so that they can be augmented or replaced when needed. [SP 800-57-part1] and [SP 800-131A] specify NIST-recommended lifetimes of government-approved cryptographic algorithms. A CKMS should only use algorithms whose security lifetime will cover the anticipated lifetime of the CKMS and the information that it protects. If the CKMS is intended to remain in service beyond the security lifetimes of its cryptographic algorithms and key lengths, then there should be a transition strategy for migration to stronger algorithms and key lengths in the future.

When transitioning from one cryptographic algorithm to another, a smooth transition often requires the capability to support the use of at least two algorithms (perhaps with different key lengths) simultaneously so that interoperability can be maintained until all participants have the capability to operate with the new algorithm. In this case, the cryptographic protocols should be designed to identify and negotiate which algorithm will be used in a particular key establishment transaction. It should also be noted that the security of data protected by different algorithms at different times is no greater than the weakest algorithm. Therefore, it may be best to transition as quickly as feasible.

FR:7.1 The CKMS design **shall** specify how interoperability requirements across device interfaces are to be satisfied.

FR:7.2 The CKMS design **shall** specify the standards, protocols, interfaces, supporting services, commands and data formats required to interoperate with the applications it is intended to support.

FR:7.3 The CKMS design **shall** specify the standards, protocols, interfaces, supporting services, commands and data formats required to interoperate with other CKMS for which interoperability is intended.

FR:7.4 The CKMS design **shall** specify all external interfaces to applications and other CKMS.

FR:7.5 The CKMS design **shall** specify all provisions for transitions to new, interoperable, peer devices.

FR:7.6 The CKMS design **shall** specify any provisions provided for upgrading or replacing its cryptographic algorithms.

FR:7.7 The CKMS design **shall** specify how interoperability will be supported during cryptographic algorithm transition periods.

FR:7.8 The CKMS design **shall** specify its protocols for negotiating the use of cryptographic algorithms and key lengths.

8. Security Controls

A CKMS requires security controls to protect its components and devices, along with the data that they contain. For example:

a) A CKMS should be physically protected so that its components, devices, and the sensitive data contained within the CKMS are protected from unauthorized disclosure and modification.

b) A CKMS will likely require computer systems to perform functions, such as key generation, key storage, key recovery, key distribution, cryptographic module control, and metadata management. Controls should exist to ensure that these functions are correctly performed.

c) A CKMS will likely be networked to distribute keys and metadata to users and other end entities. In such situations, the CKMS should be protected using network security control devices.

d) A CKMS should generate, store, use and protect cryptographic keys using a cryptographic module.

e) A CKMS should apply necessary cryptographic protections to keys before they are output from a cryptographic module.

The following subsections of this section describe requirements for each of these types of security controls.

8.1 Physical Security Controls

CKMS components and devices should be physically protected in order to ensure information security. Without good physical security, the components and devices could be tampered with and the hardware and/or software could be modified to bypass security.

A CKMS may include facilities that provide third-party key management services, such as a Certification Authority, Key Distribution Center, Registration Authority, or Certificate Directory and also end-to-end communication devices, such as personal computers, personal digital assistants, smart phones, and intelligent sensing devices.

A CKMS may include one or more primary facilities and one or more backup facilities that provide disaster recovery capabilities. Each of these facilities should have physical protection, either by controlling access to the entire facility or by controlling access to the sensitive components or devices within the facility. The use of backup systems for disaster recovery is discussed in Section 10.4.

One or more of the following mechanisms should be chosen to physically protect a CKMS facility, depending on the security criticality of its components and devices. The following are examples of physical security mechanisms. Some of the mechanisms listed below are detection mechanisms that should be augmented with appropriate prevention mechanisms.

a) Fences,
b) Gates, doors, and covers,
c) Guards,
d) Locks (keyed or combination),
e) Tamper detection and protection,
f) Passwords
g) Badges
h) Card and token systems,
i) Biometric devices,
j) Alarm systems,
k) Surveillance camera,
l) Audit systems, and
m) Entry and exit logs.

FR:8.1 The CKMS design **shall** specify each of its CKMS devices and their intended purposes.

FR:8.2 The CKMS design **shall** specify the physical security controls for protecting each device containing CKMS components.

8.2 Operating System and Device Security Controls

This section addresses the computer security controls for operating systems and CKMS devices. Note that the devices of a CKMS that incorporate a general-purpose operating system should also have computer security controls.

8.2.1 Operating System Security

A secure operating system is the foundation of a secure computer system. Without a secure operating system, the security of the programs and data on the computer system cannot be assured. A secure operating system has the following security features:

i.   BIOS protection features to ensure the proper instantiation of the operating system at start-up (see [SP 800-147]).

ii.   Self-protection features to protect the operating system from unauthorized modification by users and user processes;

iii. Isolation features to provide and maintain separate domains of execution for the users and user processes so that they do not interfere with each other and thus compromise a security policy requirement for data separation;

iv.   Access controls and operating system features that allow users to share data based on user, group or other metadata elements;

v.   Event-logging capabilities in order to support personal accountability and to investigate anomalies; and

vi.   User CKMS account management, including entity identification and authentication.

A secure operating system depends on a trusted hardware platform running secure code. The trusted hardware platform often enforces two or more states in order to provide privileged operations, such as memory and I/O management.

In some situations, a secure operating system is an isolation kernel (also known as hypervisor), which provides virtual machines to the guest operating systems and CKMS applications running on top of the guest operating systems. In this architecture, the isolation kernel views the guest operating systems as the applications.

CKMS devices that perform dedicated security functions and are not built with general-purpose capabilities can have reduced or minimal operating system requirements. As an example, consider a special-purpose device loaded with firmware and/or software to perform intrusion detection functions. This device may not have an operating system, and hence, has no operating system security requirements. Another example is a firewall or intrusion detection system built on a "locked-down" (i.e., non-modifiable) operating system so that the capability to load additional code is not available.

FR:8.3 The CKMS design **shall** specify all secure operating system requirements (including any required operating system configurations) for each CKMS device.

FR:8.4 The CKMS design **shall** specify which of the following hardening[20] features are enforced by the CKMS:
   a) Removing all non-essential software programs and utilities from the computer;
   b) Using the principle of least privilege to control access to sensitive system features and applications;
   c) Using the principle of least privilege to control access to sensitive system and application files and data;
   d) Limiting user accounts to those needed for legitimate operations, i.e., disabling or deleting the accounts that are no longer required;
   e) Running the applications with the principle of least privilege;
   f) Replacing all default passwords and keys with strong passwords and randomly generated keys, respectively;
   g) Disabling or removing network services that are not required for the operation of the system;
   h) Disabling or removing all other services that are not required for the operation of the system;
   i) Disabling removable media, or disabling automatic run features on removable media and enabling automatic malware checks upon media introduction;
   j) Disabling network ports that are not required for the system operation;
   k) Enabling optional security features as appropriate; and

---

[20] Hardening is further discussed in Section 11.4.

l) Selecting other configuration options that are secure.

FR:8.5: The CKMS design **shall** specify the BIOS protection features that ensure the proper instantiation of the operating system.

8.2.2   Individual CKMS Device Security

A CKMS may consist of a variety of devices. It is preferable that each device be designed to protect itself from unauthorized use. Otherwise, externally applied protections are necessary. Depending on the system design and functional requirements, a CKMS device may provide finer-grained access control and device-specific event logging that is not captured by the host operating system. For example, a cryptographic module could have its own access control system. Thus, a well-designed CKMS device should have the following security features:

a) Self-protection from other CKMS devices (e.g., by utilizing operating system process isolation),

b) Self-protection from CKMS device users,

c) Isolation features to provide and maintain separate sessions for the users and user processes so that they do not interfere with each other and thus violate the security policy of data separation,

d) Fine-grained access controls on CKMS device-level objects (e.g., keys and metadata or Data Base Management System rows and tables) to allow users to share data based on user, group or other metadata elements,

e) CKMS device-level event logging in order to support personal accountability and to investigate anomalies, and

f) User account management for the CKMS.

FR:8.6 The CKMS design **shall** specify the security controls required for each CKMS device.

FR:8.7 The CKMS design **shall** specify the device/CKMS secure configuration requirements and guidelines that the hardening is based upon.

8.2.3   Malware Protection

CKMS devices that receive communications, data, files, etc. over unprotected networks should scan the information for malware. Malware protection may be less critical if no information is received over unprotected networks, or if all information is strongly (e.g., cryptographically) authenticated. Malware protection falls into the following three general categories:

a) Anti-virus software that protects CKMS devices from unwittingly installing and executing programs that perform unintended actions and may cause a security compromise,

b) Anti-spyware software that protects CKMS devices from unauthorized parties obtaining system administrator status or authorized user status, and prevents the spyware from taking on authorized device behavior, and

c) Rootkit detection and prevention software that protects CKMS devices from rootkit malware that changes the configuration setting of the operating system in order to replace system code and hide processes and files, including the rootkit code itself, from anti-virus and anti-spyware software.

The integrity of operating system and CKMS application software should be checked upon installation and periodically thereafter. Examples of software integrity verification upon installation include the chain of custody for the software and the verification of integrity codes (e.g., hash values, message authentication codes, and digital signatures) used to assure that the software has not been modified. Examples of periodic verification include the daily verification of hash values, message authentication codes, digital signatures, and modification dates on the installed software, etc.

In order to be effective, malware protection should be configured to perform the following:

a) A daily scan of installed software,

b) A scan of removable media when first introduced into the CKMS,

c) A scan of newly installed software and data files,

d) A weekly update of the malware protection software, and

e) A weekly update of the malware signature database.

FR:8.8 The CKMS design **shall** specify the following malware protection capabilities for CKMS devices:
  a) Anti-virus protection software, including the specified time periods and events that trigger anti-virus scans, software update, and virus signature database updates;
  b) Anti-spyware protection software, including the specified time periods and events that trigger anti-spyware scans, software update, and virus signature updates; and
  c) Rootkit detection and protection software, including the specified time periods and events that trigger rootkit detection, software update, and signature updates.

FR:8.9 The CKMS design **shall** specify the following software integrity check information for operating system and CKMS application software:
  a) If software integrity is verified upon installation, indicate how the verification is performed; and
  b) If software integrity is verified periodically, indicate how often the verification is performed.

8.2.4 Auditing and Remote Monitoring

A CKMS should audit security-relevant events by detecting and recording the event, the date and time of the event, and the identity or role of the entity initiating the event. The audit log should provide a record of the relevant security functions performed. The audit capability may be spread over several CKMS devices and locations. The audit capability should also have the ability to detect and report to the audit administrator role any unusual events that should be investigated as soon as possible. The audit capability and audit log should be protected from unauthorized modification so that the integrity of the audit system can be assured.

Automated assessment tools, such as those specified in the Security Content Automation Protocol (SCAP), are becoming increasingly useful in assessing the current status and integrity of computer systems. These tools can interrogate an operating system to determine its status in real time (see [SP 800-126]). Software version numbers can be checked for currency, and the integrity and confidentiality of the data files can be verified. Monitoring tools may execute on the platform being monitored or on another platform dedicated to monitoring other hosts. These monitoring tools can detect modifications to system files or their access control attributes and post alerts and audit events (see Section 6.8.5).

FR:8.10 The CKMS design **shall** specify the auditable events supported and indicate whether each event is fixed or selectable.

FR:8.11 For each selectable, auditable event, the CKMS design **shall** specify the role(s) that has the capability to select the event.

FR:8.12 For each auditable event, the CKMS design **shall** specify the data to be recorded[21].

FR:8.13 The CKMS design **shall** specify what automated tools are provided to assess the correct operation and security of the CKMS.

FR:8.14 The CKMS design **shall** specify system-monitoring requirements for sensitive system files to detect and/or prevent their modification or any modification to their security attributes, such as their access control lists.

8.3 Network Security Control Mechanisms

This section addresses the network security control mechanisms for each of the computer systems involved in the CKMS. Examples of network security control mechanisms include:
 a) Firewalls,
 b) Filtering Routers,
 c) Virtual Private Networks (VPNs),
 d) Intrusion Detection Systems (IDS),
 e) Intrusion Prevention Systems (IPS),
 f) Adaptive Network Security Controls,

---

[21] Examples of recorded data include the unique event identifier, the date and time of the event, the subject (e.g., user, role or software process) causing the event, the success or failure of the event and the event-specific data.

     i.   Adaptive Filtering mechanisms,
    ii.   Adaptive Detection mechanisms, and
   iii.   Adaptive Prevention mechanisms.

Networked CKMS devices should be protected using a combination of firewalls and intrusion detection and prevention systems. While firewalls provide protection by filtering out unwanted and potentially dangerous protocols and by examining permitted protocol data to reduce the chances of a successful attack, intrusion detection and prevention systems complement firewalls by examining host and network activity to determine if the systems are being attacked and by preventing the detected attacks. Thus, both firewall and intrusion detection and prevention systems should be used.

Boundary control devices (such as firewalls, filtering routers, VPNs, IDS, and IPS) could be hosted on computer systems (see Section 8.2) or could be implemented in dedicated hardware. These devices should be placed in physically secure locations (see Section 8.1 for physical security controls) and should only be configured with user accounts and network services that are required for secure operation. In order to provide defense-in-depth, boundary control functions should also be implemented directly in CKMS devices. Such controls are termed "host-based" firewalls.

FR:8.15 The CKMS design **shall** specify the boundary protection mechanisms employed by the CKMS.

FR:8.16 The CKMS design **shall** specify:
  a) The types of firewalls used and the protocols permitted through the firewalls, including the source and destination for each type of protocol; and
  b) The types of intrusion detection and prevention systems used, including their logging and security breach reaction capabilities.

FR:8.17 The CKMS design **shall** specify the methods used to protect the CKMS devices against denial of service.

FR:8.18 The CKMS design **shall** specify how each method used protects against the denial of service.

8.4 Cryptographic Module Controls

A cryptographic module is a set of hardware, software and/or firmware that implements cryptographic-based security functions (e.g. cryptographic algorithms and key establishment schemes). The module encompasses everything within its cryptographic boundary[22] and includes the boundary itself. Each cryptographic module should be built in accordance with and to enforce a cryptographic module security policy (e.g. see [FIPS 140]).

---

[22] A cryptographic boundary is an explicitly defined perimeter that establishes the boundary of all components of a cryptographic module.

Two primary security issues should be addressed regarding the security of the contents of cryptographic modules: the integrity of the security functions and the protection of the cryptographic keys and metadata. For example, [FIPS 140] specifies requirements on cryptographic modules for protecting keys within the module and maintaining the integrity of the module's security functions. Techniques such as the *software/firmware integrity test* and *known-answer test*, along with physical protection from unauthorized access and/or alteration, are specified in the FIPS. Since the cryptographic keys are present in plaintext form for some period of time within the module, physical security measures are necessary to protect keys from unauthorized disclosure, modification, and substitution. A cryptographic module may provide the necessary physical protection. Otherwise, a larger, physically protected space that includes the module is needed.

Vendors of hardware cryptographic products or modules often build physical security safeguards into their devices by using strong metal cases, locks, alarms, and key destruction mechanisms. However, software cryptographic applications running on general-purpose computers face additional risks because these computers were not designed and built to provide sufficient protection for cryptographic keys. In fact, the very computers on which the cryptography runs usually contain software written by individuals that have not been vetted for security. It is, therefore, critical that cryptographic software running on a general-purpose computer is both physically protected (i.e., kept in a safe environment) and logically protected from exploitation by distrusted (perhaps hostile) code. [FIPS 140] provides guidance regarding these protections.

FR:8.19 The CKMS design **shall** identify the cryptographic modules that it uses and their respective security policies, including:
   a) The embodiment of each module (software, firmware, hardware, or hybrid),
   b) The mechanisms used to protect the integrity of each module,
   c) The physical and logical mechanisms used to protect each module's cryptographic keys, and
   d) The third-party testing and validation that was performed on each module (including the security functions) and the protective measures employed by each module.

9. Testing and System Assurances

A CKMS device should undergo several types of testing to ensure that it has been built to conform to its design, that it conforms to certain standards, that it continues to operate according to its design, that it does not perform functions that may compromise its security, that it is interoperable with other CKMS devices, and that it can be used in the larger systems for which it is intended with reasonable assurance of preserving security.

Since testing is restricted to a finite number of cases that is typically much less than the total number of possibilities, testing does not guarantee that a device or system is correct or secure in all cases. Thus, the value of passing a test suite is directly related to the comprehensiveness and representation of the selected test cases.

A CKMS device may undergo tests in the categories listed below.

9.1 Vendor Testing

Device vendors test their devices to detect and eliminate errors and then to verify that they work as expected. The techniques and specifics of this category of testing are often considered proprietary information by the vendor and are generally not made public.

FR:9.1 A CKMS design **shall** specify the non-proprietary vendor testing that was performed on the system and passed.

9.2 Third-Party Testing

A vendor or customer may request that a third-party test a CKMS device for conformance to a particular standard. Third-party testing provides confidence that the vendor did not overlook some flaw in its own testing procedures. For example, the National Institute of Standards and Technology has established several programs for validating the conformance of products to its cryptographic standards and recommendations.

FR:9.2 The CKMS design **shall** specify all third-party testing programs that have been passed to date by the CKMS or its devices.

9.3 Interoperability Testing

Interoperability testing, in its most general form, merely tests that two or more devices can be interconnected and operate with one another. This means that the data exchanged between the devices is in a format that each device can process. Interoperable devices may be interconnected to form a system, and interoperable systems may be interconnected to form a network. Note that this type of testing does not necessarily test the internal functioning of the individual device. If a device performs a unique function, interoperability testing may not verify the correct operation of that function.

Another form of interoperability testing is used to verify that a device (i.e., the device-under-test) appears to be working properly. If another device that performs the same or complementary functions (i.e., the assured-baseline device) has been tested and verified to operate correctly, the device-under-test may be tested to verify that it interoperates with the assured-baseline device; this provides some assurance that the device-under-test operates correctly. For example, a device performing key establishment could be tested against another such device that is believed to operate correctly. If the two devices agree on the established key, then the test is passed. This testing produces more credible results when the device-under-test and the assured-baseline device are independently designed and built by different organizations or by individuals working independently of those involved with designing and implementing the device-under-test. This is because two devices built by the same group may interoperate consistently, but incorrectly, with each other. The NIST Cryptographic Algorithm Validation Program (CAVP) performs interoperability testing on implementations of NIST-approved cryptographic algorithms using implementations developed by NIST as assured-baseline implementations.

FR:9.3 If a CKMS claims interoperability with another system, then the CKMS design **shall** specify the tests that have been performed and passed that verify the claim.

FR:9.4 If a CKMS claims interoperability with another system, then the CKMS design **shall** specify any configuration settings that are required for interoperability.

9.4 Self-Testing

A device may be designed, implemented, and operated correctly when first deployed, but fail some time later. When a failure is detected in a device, the device can be repaired or replaced, but undetected failures can have major security implications. A CKMS should use devices that test themselves for integrity and security failures. For example, [FIPS 140] specifies several self-tests to help verify the correct operation of a cryptographic module, including all its security functions.

FR:9.5 The CKMS design **shall** specify all self-tests created and implemented by the designer and the corresponding CKMS functions whose correct operation they verify.

9.5 Scalability Testing

Scalability is the ability of a system, network, or process, to correctly process increasing amounts of work in a graceful manner, or its ability to be enlarged to accommodate that increase. Scalability testing involves testing a device or system to learn how it reacts when the number of transactions to be processed or participants to be handled over a given period of time increases dramatically. Every device has its limitations, but some device designs scale better than others. If systems are not designed for modular scalability, adding additional devices may not be feasible. In addition, subtle problems often arise that cannot be solved by simply buying more equipment. Scalability testing is used to stress devices and systems so that these problems are known and mitigated before they become fully operational.

FR:9.6 The CKMS design **shall** specify all scalability analysis and testing performed on the system to date.

9.6 Functional Testing and Security Testing

The types of tests previously described can be performed to meet particular test goals. Functional testing attempts to verify that an implementation of some function operates correctly. A functional test might determine that a cryptographic algorithm implementation correctly computes the ciphertext from the plaintext, given the key. Security testing attempts to verify that an implementation functions securely. A security test might determine that, while a cryptographic algorithm implementation functions correctly (i.e. it produces the correct results), fluctuations in power consumption during the cryptographic process could lead to the compromise of the key. Thus, a cryptographic algorithm implementation could pass functional testing, but fail security testing.

Penetration testing is a specific type of security testing in which a team of penetration-testing experts develops penetration scenarios for the system as a whole and then evaluates the risk of a successful penetration. Note that individual product/device penetration testing may be conducted as part of the CKMS security assessment (see Section 11). The scope of penetration testing should include personnel, facilities, and procedures. The penetration team attempts to bypass the security safeguards with the goal of defeating CKMS security. Any findings made by the penetration testing team should be addressed before initial deployment.

FR:9.7 The CKMS design **shall** specify the functional and security testing that was performed on the system and the results of the tests.

9.7 Environmental Testing

CKMS designs often assume a particular environment (e.g., temperature range and voltage range) for their devices or systems. The CKMS devices or systems are then built for that environment and tested within that environment. If the device or system is used in a different environment, secure operation could be lost. Military systems are often ruggedized to handle the extreme conditions under which they may be used. This extra protection frequently comes at an additional cost.

FR:9.8 The CKMS design **shall** specify the environmental conditions in which the CKMS is designed to be used.

FR:9.9 The CKMS design **shall** specify the results of environmental testing that was performed on the CKMS devices, including the results of all tests stressing the devices beyond the conditions for which they were designed.

9.8 Development, Delivery, and Maintenance Assurances

The secure development, delivery, and maintenance of CKMS products can play a significant role in the security of the CKMS. The following areas should be considered:
   a) Configuration Management,
   b) Secure Delivery,
   c) Development and Maintenance Environmental Security, and
   d) Flaw Remediation.

Each of these areas is described in the following subsections.

9.8.1 Configuration Management

A CKMS should incorporate products that are developed and maintained under an appropriate configuration management system in order to ensure that security is not reduced and functional flaws are not introduced due to unauthorized or unintentional changes to the products.

FR:9.10 The CKMS design **shall** specify:

a) The devices (including their source code, documentation, build scripts, executable code, firmware, hardware, documentation, and test code) to be kept under configuration control.

b) The protection requirements (e.g., formal authorizations and proper record keeping) to ensure that only authorized changes are made to the components and devices under configuration control.

9.8.2 Secure Delivery

When products to be used in a CKMS are delivered, assurance of secure delivery (i.e. that the products received are the exact products that were ordered) is necessary.

FR:9.11 The CKMS design **shall** specify secure delivery requirements for the products used in the CKMS, including:

a) Protection requirements to ensure that the product has not been tampered with during the delivery process or that tampering is detected,

b) Protection requirements to ensure that the product has not been replaced during the delivery process or that replacement is detected,

c) Protection requirements to ensure that an unrequested delivery is detected, and

d) Protection requirements to ensure that the product delivery is not suppressed or delayed and that suppression or delay is detected.

9.8.3 Development and Maintenance Environmental Security

The CKMS development and maintenance environments must be properly protected from physical, personnel, and IT hacking threats. Tools such as compilers, software linkers, and text editors should not be automatically trusted.

FR:9.12 The CKMS design **shall** specify the security requirements for the development and maintenance environments of the CKMS, including:

a) Physical security requirements,

b) Personnel security requirements, such as clearances and background checks for developers, testers, and maintainers,

c) Procedural security, such as multi-person control and separation of duties,

d) Computer security controls to protect the development and maintenance environment and to provide access control to permit authorized user access,

e) Network security controls to protect the development and maintenance environment from hacking attempts,

f) Cryptographic security control to protect the integrity of software and its control data under development, and

g) The means used to ensure that the tools (e.g., editors, compiler, software linkers, loaders, etc.) are trustworthy and are not sources of malware.

9.8.4   Flaw Remediation Capabilities

A CKMS should have the capability to detect, report, and fix flaws in an expeditious and secure manner. A CKMS that employs automated techniques is highly desirable because it can continuously monitor its own security status, report potential problems to an authorized person fulfilling an appropriate CKMS role, and minimize reliance on human monitoring of events that occur infrequently.

FR:9.13 The CKMS design **shall** specify the CKMS capabilities for detecting system flaws, including:
  a) Known-answer tests,
  b) Error detection codes,
  c) Anomaly diagnostics, and
  d) Functional Testing.

FR:9.14 The CKMS design **shall** specify the CKMS capability for reporting flaws, including: the capability to produce status report messages with confidentiality, integrity and source authentication protections, and to detect unauthorized delays.

FR:9.15 The CKMS design **shall** specify the CKMS capability for analyzing flaws and creating/obtaining fixes for likely or commonly known flaws.

FR:9.16 The CKMS design **shall** specify its capability to transmit fixes with confidentiality, integrity and source authentication protections and to detect unauthorized delays.

FR:9.17 The CKMS design **shall** specify its capability for implementing fixes in a timely manner.

10. Disaster Recovery

The use of a CKMS to manage cryptographic keys and metadata that are used to protect information has the additional risk that a failure of the CKMS may hamper or prevent access to the information. For example, the failure of the decrypting capability may delay or prevent the use of enciphered data. This section describes how operational continuity can be achieved in the event of component failures or the corruption of keys and metadata

10.1 Facility Damage

A CKMS should be located in physically secure and environmentally protected facilities. In addition, the CKMS management should provide for backup and recovery in the event that damage to the CKMS occurs. The backup and recovery facilities should be designed, implemented, and operated at a level that is commensurate with the value and sensitivity of the data and the operations being protected. When a CKMS facility is damaged, operations should be transferred to a backup facility, and keys that could have been disclosed accidentally should be immediately placed on Compromised Key or Certificate

Revocation Lists and replaced, if appropriate. Wind and water damage are the common environmental risks; fire is both an environmental risk and a facility design-dependent risk.

FR:10.1 The CKMS design **shall** specify the required environmental, fire, and physical access control protection mechanisms and procedures for recovery from damage to the primary and all backup facilities.

10.2 Utility Service Outage

A CKMS requires reliable utility services, including electricity, water, sewer, air conditioning, heat, and clean air in order to assure the continued availability of the CKMS. Electrical power sufficient to satisfy the requirements of all electronic devices, as well as human safety and comfort provisions in normal operations and during emergencies, should be available in the primary and all backup CKMS facilities.

FR: 10.2 The CKMS design **shall** specify the minimum as well as recommended electrical, water, sanitary, heating, cooling, and air filtering requirements for the primary and all backup facilities.

10.3 Communication and Computation Outage

A CKMS needs sufficient communication and computation capability to perform its required functions and to provide the services required by its users. Long-distance communication capabilities are typically offered by commercial vendors, and many computer vendors can provide computers sufficient for large key-management applications. Redundant communications equipment is often installed in a CKMS to assure high availability. Remote on-line backup facilities can be used to provide even higher service availability, especially against potential environmental (e.g., weather) risks. The ability to quickly access alternative communications services is highly desirable in the event of a communications service failure.

FR:10.3 The CKMS design **shall** specify the communications and computation redundancy present in the design and required to be available during operation in order to assure continued operation of services commensurate with the anticipated needs of users, enterprises, and CKMS applications.

10.4 System Hardware Failure

Since the CKMS is critical for the secure operation of the information management system that it supports, it is desirable to minimize the impact of hardware failures of CKMS components and devices. Several approaches to recover from hardware failure exist. These approaches tend to trade-off the ease and speed of recovery with cost. The redundancy of backup systems can provide assurance that the operational impact of a single hardware failure is quickly detected and that a fully operational secure state is quickly attained. In order for backup systems to be most effective, they should maintain synchronization with the primary system. Backup systems that continuously maintain synchronization with the primary system are called "hot" backups. These systems are capable of immediately taking over the responsibilities of the primary system. Some systems synchronize periodically and have a catch-up procedure to bring the backup system up to the state that the primary system had just before the failure occurred.

It is essential that backup systems have as much independence from the primary system as possible so that a failure to the primary system does not also result in the same failure to the backup. For example, a power surge on a common power line could cause both the primary and its backup to fail. In order to maximize independence, it is best not to co-locate a backup system with the primary system. Multiple backup systems can be used to provide error detection capabilities. For example, if three systems are all performing the same functions, then the failure of any one system can be detected and corrected by taking the majority vote of the three systems as the valid result. Since redundancies increase the cost of providing services, system vendors and CKMS owners strive to find an optimum trade-off between redundancy and cost.

FR:10.4 The CKMS design **shall** specify the strategy for backup and recovery from failures of hardware components and devices.

10.5 System Software Failure

Software failures are typically of two types:

    a. Software that was incorrectly written so that it does not perform as desired when a particular condition occurs, and

    b. Software that was written correctly, but has been garbled when it resides in memory before it is executed.

Many software failures can be avoided by writing code using good, well-established programming practices. Several books have been written on this topic that involve the use of good programming procedures, addressing boundary conditions, protecting against memory overflows, code analysis, and proper software testing.

Failures that garble code should be detected as soon as possible. This may be accomplished by validating an error detection code or known-answer test on the software before it is run. If an error is detected, the program can be implemented to enter an error state and output an error indicator. This permits the error to be detected and repaired before the code is used operationally. These tests can also be executed periodically as desired. Redundant systems, as previously discussed in Section 10.4, can also be used to recover from this type of failure.

When the CKMS in a primary facility is recovered to a known secure state, some of the data created since the last secure state may be lost. A CKMS should be implemented and operated under the assumption that a catastrophe will eventually occur. Therefore, it is recommended that full secure-state system backups be made on a regular basis, and that the latest CKMS secure state be reloaded into a repaired and ready CKMS.

FR:10.5 The CKMS design **shall** specify all techniques provided by the CKMS to verify the correctness of the system software.

FR:10.6 The CKMS design **shall** specify all techniques provided by the CKMS to detect alterations or garbles to the software once it is loaded into memory.

FR:10.7 The CKMS design **shall** specify the strategy for backup and recovery from a major software failure.

10.6 Cryptographic Module Failure

Cryptographic modules should have built-in tests that are adequate to detect hardware, software, or firmware failures. Cryptographic modules may have pre-operational, conditional, and periodic self-tests. For example, when a failure is detected within a [FIPS 140]-2 validated module, control is passed to an error state that outputs an error indicator and determines whether the error is a non-recoverable type (i.e. one that requires service, repair, or replacement) or a recoverable type (i.e., one that requires initialization or resetting). In most cases, sensitive data should not be output from the module while it is in the error state. If the error is recoverable, the module should be rebooted and pass all power-up self-tests before continuing normal processing. If the error recurs on repeated attempts to reboot, then the module should be replaced.

FR:10.8 The CKMS design **shall** specify what self-tests are used by each cryptographic module to detect errors and verify the integrity of the module.

FR:10.9 The CKMS design **shall** specify how each cryptographic module responds to detected errors.

FR:10.10 The CKMS design **shall** specify its strategy for the repair or replacement of failed cryptographic modules.

10.7 Corruption of Keys and Metadata

Cryptographic keys and metadata may be corrupted in transmission or in storage. Corrupted keys and metadata should be replaced or corrected as soon as the corruption is detected. The replacement of corrupted keys and metadata typically involves the establishment or storage of a new key and metadata. If a corrupted key or a key with corrupted metadata was used to protect data, the security consequences should be evaluated, since a loss or compromise of sensitive data could result. Key establishment and key storage protocols are frequently designed to detect and replace corrupted keys.

A major disaster could result in large numbers of operational keys and metadata being lost or corrupted beyond recovery in primary storage. If a key recovery, backup, or archive system exists, then the keys and metadata can and should be restored. However, if the keys were not backed-up or archived, then they would have to be replaced with new keys, and the information that the original keys protected could be lost.

FR:10.11 The CKMS design **shall** specify its procedures for backing-up and archiving cryptographic keys and their metadata.

FR:10.12 The CKMS design **shall** specify its procedures for restoring or replacing corrupted keys and metadata that have been stored or transmitted.

11. Security Assessment

CKMS security may be assessed at any time throughout the lifetime of the CKMS. This section highlights assessment considerations to be made during the initial deployment, during periodic (e.g., annual) reviews, and during incremental assessments after major changes. For additional information on U.S. Government security assessment practices, see [SP 800-37], [SP 800-53], [SP 800-53A], and [SP 800-115].

11.1 Full Security Assessment

Prior to or upon deployment of a CKMS, its security should be fully assessed. The activities that can be undertaken to assess the security of the CKMS include the following:
   a) Review of third-party validations,
   b) Architectural review of the system design,
   c) Functional and security testing of the CKMS, and
   d) Penetration testing of the CKMS.

Each of these activities is described in the following subsections.

FR:11.1 The CKMS design **shall** specify the necessary assurance activities to be undertaken prior to or in conjunction with a full CKMS security assessment.

FR:11.2 The CKMS design **shall** specify the circumstances under which a full security assessment is repeated.

11.1.1 Review of Third-Party Validations

While there are currently no formal validation programs for the security of a CKMS, the following validation programs exist for certain devices of a CKMS:
   a) NIST-approved cryptographic algorithm implementations can be validated under the NIST Cryptographic Algorithm Validation Program (CAVP),
   b) Cryptographic modules can be validated for conformance to [FIPS 140]-2 under the NIST Cryptographic Module Validation Program (CMVP),
   c) Non-cryptographic security and hardware (e.g. operating systems, DBMS, or firewall) can be validated using the Common Criteria Standard (see [ISO/IEC 15408 Parts 1-3]) under the National Information Assurance Partnership (NIAP), and
   d) A CKMS, or parts thereof, could also be validated by a private entity hired by the vendor or a sponsor.

While these validation programs do not guarantee security, they can significantly increase confidence in the security and integrity of the CKMS.

FR:11.3 The CKMS design **shall** specify all validation programs under which any of the CKMS devices have been validated.

FR:11.4 The CKMS design **shall** specify all validation certificate numbers for its validated devices.

11.1.2 Architectural Review of System Design

Under this activity, a team of experts is assembled to evaluate the CKMS architecture. The architecture review team should have access to the CKMS design information, the third-party validation information, and the results of all available CKMS testing. The recommendations provided by the architecture review team should be reviewed by the designer, and the recommendations that are selected to be integrated into the CKMS should result in documented and implemented design changes. The architecture review team should also make recommendations for penetration-testing scenarios that are reviewed by the CKMS management. Penetration tests that are selected to be performed should be done in a timely manner in accordance with CKMS management direction.

The architecture review team should have expertise in cryptography, cryptographic protocols, secure system design, network security, computer security, human usability/accessibility, functional safety, distributed system design, and applicable information system law and regulations.

FR:11.5 The CKMS design **shall** specify whether an architectural review is required as part of the full security assessment.

FR:11.6 If an architectural review is required, then the CKMS design **shall** specify the skill set required by the architectural review team.

11.1.3 Functional and Security Testing

Functional and security testing is typically performed as part of the full security assessment, the periodic security review, and the incremental security assessment. A variety of functional and security tests may be performed by the vendor, the information owner, or a trusted third-party (see Section 9).

FR:11.7 The CKMS design **shall** specify all required functional and security testing of the CKMS.

FR:11.8 The CKMS design **shall** report the results of all functional and security tests performed to date.

11.1.4 Penetration Testing

Penetration testing tests the CKMS to verify the extent to which it resists active attempts to compromise its security. This type of testing requires security experts who are knowledgeable about the typical attack techniques and system weaknesses, and who also have the ability to create and try new or unsuspected attack methods. The attack group should contain some individuals who are not part of the CKMS design team and who do not have preconceived notions about its security. Successful attack methods often involve using the system in unintended or unsuspected ways.

FR:11.9 The CKMS design **shall** specify the results of any completed penetration testing performed to date.

11.2 Periodic Security Review

This review consists of an examination of the system controls, physical controls, procedural controls and personnel controls to ensure that these controls are in place and operational as claimed. Changes to the system since the previous security review should be examined to ensure that the products/devices are operating with the latest updates and security patches in secure configurations, and that the products continue to maintain their third-party security ratings. Issues identified as the result of the review should be addressed. In addition, periodic functional and security testing should be performed (see Section 9.6).

FR:11.10 The CKMS design **shall** specify the periodicity of security reviews.

FR:11.11 The CKMS design **shall** specify the scope of the security review in terms of the CKMS devices.

FR:11.12 The CKMS design **shall** specify the scope of the periodic security review in terms of the activities undertaken for each CKMS device under review.

FR:11.13 The CKMS design **shall** specify the functional and security testing to be performed as part of the periodic security review.

11.3 Incremental Security Assessment

When the system has undergone significant changes, an incremental assessment of the changes in the following areas described in Section 11.1 should be performed:
   a) Changes to third-party-validated devices since the previous security assessment,
   b) Architecture review of the system design changes, and
   c) Functional and security testing of the CKMS.

If the cumulative system changes are significant, a full CKMS security assessment as specified in Section 11.1 should be conducted.

FR:11.14 The CKMS design **shall** specify the circumstances under which an incremental security assessment should be conducted.

FR:11.15 The CKMS design **shall** specify the scope of incremental security assessments.

11.4 Security Maintenance

While a CKMS may be designed, developed and deployed to provide a specific level of protection (e.g., low, medium, or high), the protection provided may be reduced as configuration changes are made and as new threats are found. In order to maintain and enhance the security of the system, the CKMS should be properly upgraded, reviewed and tested against hardening guidelines. Examples of hardening activities include updating the CKMS with the latest security patches, periodic review of the system configuration against the hardening guidelines, periodic testing of the CKMS against hardening guidelines, application of revised hardening guidelines, and periodic penetration testing.

FR:11.16 The CKMS design **shall** list the hardening activities required to be performed in order to maintain its security.

12. Technological Challenges

A CKMS should be designed and implemented to have a security lifetime of many years. Therefore, the designer should consider possible threats resulting from advances in technology that may render the CKMS insecure. Some examples are discussed below.

a) New Attacks on Cryptographic Algorithms
   A cryptographic algorithm has an expected security life. However, as time passes, new attacks may be found that reduce its security life. This, in turn, is likely to reduce the security lifetime of the CKMS that relies on the algorithm to protect its keys and metadata. Eventually, the cryptographic algorithm may need to be upgraded or replaced altogether.

   Cryptographic algorithms should be implemented within cryptographic modules in a manner so that the algorithms can be replaced or updated without significantly affecting the rest of the implementation. For example, key lengths and block lengths should be variable so that they may be easily increased if and when necessary.

b) New Attacks on Key Establishment Protocols
   Weaknesses are often found in protocols after they have been in use for several years. Protocols are seldom evaluated to the same extent as cryptographic algorithms, and it is often difficult to upgrade a protocol once it is widely used. Therefore, it is important that a CKMS uses evaluated and tested protocols for key establishment.

c) New Attacks on CKMS Devices
   New methods for logically attacking and subverting computer-based systems are continuously being discovered. The CKMS designer should prevent, to the

maximum extent that is feasible, external access to CKMS devices by unauthorized parties. The access control mechanisms upon which the CKMS relies for its security should be periodically reviewed against the most recent attacks being perpetrated and upgraded as required.

d) New Computing Technologies

New computing technologies may threaten the security of a CKMS. The current threat of highest concern is that of the development of quantum computers with sufficient capability to recover cryptographic keys. The implementation of practical quantum computers could result in a major change in cryptographic security technology. For example, if large qubit quantum computers could be built, the security of integer factorization and discrete log-based public key cryptographic algorithms would be threatened. This would have a major impact on the CKMS that rely on these algorithms for the establishment of cryptographic keys. Research is currently underway to find public key algorithms that are resistant to quantum computing (e.g., lattice-based public key cryptography), but no widely accepted solution has yet been found. Research is also currently underway to find scalable, symmetric key distribution architectures that can use symmetric key algorithms (e.g., AES-256) that are currently considered more resistant to quantum computing attacks.

FR:12.1 The CKMS design **shall** specify the expected security lifetime of each cryptographic algorithm implemented in the system.

FR:12.2 The CKMS design **shall** specify which sub-functions (e.g., the hash sub-function of HMAC) of the cryptographic algorithms can be upgraded or replaced with similar, but cryptographically improved, sub-functions without negatively affecting the CKMS operation.

FR:12.3 The CKMS design **shall** specify which key establishment protocols are implemented by the system.

FR:12.4 The CKMS design **shall** specify the expected security lifetime of each key establishment protocol implemented in the system in terms of the expected security lifetimes of the cryptographic algorithms employed.

FR:12.5 The CKMS design **shall** specify the extent to which external access to CKMS devices is permitted.

FR:12.6 The CKMS design **shall** specify how all allowed external accesses to CKMS devices is controlled.

FR:12.7 The CKMS design **shall** specify the features employed to resist or mitigate the consequences of the development of new technologies, such as a quantum computing attack upon the CKMS cryptographic algorithms.

FR:12.8 The CKMS design **shall** specify the currently known consequences of a quantum computing attack upon the CKMS cryptography.

Appendix A: References

A short summary is provided for each of the items below so that the reader can immediately determine the applicability of the item to the reader's needs.

1. [FIPS 140]
   FIPS 140-2: Security Requirements for Cryptographic Modules, May 2001, www.csrc.nist.gov/publications/PubsFIPS.html.
   FIPS 140-2 specifies the requirements in eleven areas that must be met by cryptographic modules (modules) protecting U.S. Government information. This applies to hardware, software, firmware and hybrid modules. The standard provides four increasing, qualitative levels of security that are intended to cover a wide range of potential applications and environments. The security requirements cover areas related to the secure design and implementation of a cryptographic module. These areas include the cryptographic module specification; the cryptographic module ports and interfaces; the roles, services, and authentication mechanisms; the finite state model; the physical security of the module; the operational environment; cryptographic key management; electromagnetic interference/electromagnetic compatibility (EMI/EMC); self-tests; design assurance; and mitigation of other attacks. Compliance with this Standard is validated under the NIST Cryptographic Module Validation Program (CMVP).

2. [FIPS 180]
   FIPS 180-4: Secure Hash Standard (SHS), March 2012, www.csrc.nist.gov/publications/PubsFIPS.html.
   FIPS 180-4 specifies hash algorithms that can be used to generate digests of messages. The digests are used to detect whether messages have been changed since the digests were generated. Digests may be used in the generation and validation of digital signatures, random number generation and the generation of message authentication codes. Compliance with this Standard is validated under the NIST Cryptographic Algorithm Validation Program (CAVP).

3. [FIPS 186]
   FIPS 186-4: Digital Signature Standard (DSS), July 2013, www.csrc.nist.gov/publications/PubsFIPS.html.
   FIPS 186-4 specifies algorithms for applications requiring a digital signature. The standard allows the use of DSA, RSA, and ECDSA signature techniques, along with an appropriate hash function from FIPS 180-4 to compute the digital signature. FIPS 186-4 also includes requirements and algorithms for the generation of keys and domain parameters. Compliance with this Standard is validated under the NIST Cryptographic Algorithm Validation Program (CAVP).

4. [FIPS 197]
   FIPS 197: Advanced Encryption Standard (AES), November 2001,
   www.csrc.nist.gov/publications/PubsFIPS.html.
   FIPS 197 specifies a symmetric key block cipher encryption/decryption algorithm. The standard supports key sizes of 128, 192, and 256 bits and a block size of 128 bits. Compliance with this Standard is validated under the NIST Cryptographic Algorithm Validation Program (CAVP).

5. [FIPS 198]
   FIPS 198-1: The Keyed-Hash Message Authentication Code (HMAC), July 2008,
   www.csrc.nist.gov/publications/PubsFIPS.html.
   FIPS 198-1 describes a keyed-hash message authentication code (HMAC), a mechanism for message authentication using cryptographic hash functions. HMAC can be used with a NIST-approved cryptographic hash function, in combination with a shared secret key. Compliance with this Standard is validated under the NIST Cryptographic Algorithm Validation Program (CAVP).

6. [IPSEC]
   Various IPSEC RFCs under http://www.ietf.org/dyn/wg/charter/ipsecme-charter.html. These IPSEC RFCs describe how authentication, encryption, and integrity security services are provided for the IP packets. The RFCs cover the format of the security services payload for the packets, cipher suites for the security services, and key management techniques for the cryptographic algorithms used to provide the security services.

7. [ISO/IEC 15408 Parts 1-3]
   Information technology – Security techniques – Evaluation criteria for IT security,
   Part 1: Introduction and general model
   Part 2: Security functional requirements
   Part 3: Security assurance components
   http://www.iso.org/iso/catalogue.
   ISO/IEC 15408-1:2009 establishes the general concepts and principles of an IT security evaluation and specifies the general model of evaluation given by various parts of ISO/IEC 15408, which in its entirety is meant to be used as the basis for the evaluation of the security properties of IT products.

   ISO/IEC 15408-2:2005 defines the required structure and content of security functional components for the purpose of a security evaluation. It includes a catalogue of functional components that will meet the common security functionality requirements of many IT products and systems.

   ISO/IEC 15408-3:2008 defines the assurance requirements of the evaluation criteria. It includes the evaluation assurance levels that define a scale for measuring the assurance for component targets of evaluation (TOEs), the composed assurance packages that define a scale for measuring the assurance for composed TOEs, the individual assurance components from which the assurance levels and packages are composed, and the criteria for the evaluation of protection profiles and security targets.

8. [KERBEROS]
   Various Kerberos RFCs under http://www.ietf.org/dyn/wg/charter/krb-wg-charter.html.
   These KERBEROS RFCs describe how KERBEROS authentication, encryption, and ticket-granting security services are provided. The RFCs cover the format of the security services payload, cipher suites for the security services, and authentication using passwords or X.509 certificates.

9. [RFC 6960]
   Online Certificate Status Protocol (OCSP), http://www.ietf.org/rfc/rfc6960.txt.
   RFC 6960 specifies a protocol that may be used for determining the current status of a public key certificate.

10. [RFC 3161]
    Internet X.509 Public Key Infrastructure Time-Stamp Protocol (TSP),
    http://www.ietf.org/rfc/rfc3161.txt.
    RFC 3161 specifies a protocol to request and receive a trusted time stamp from a trusted third-party. The document specifies the digital signature-based structure of the time stamp token, which can be used to provide the time associated with the existence of a datum.

11. [RFC 3279]
    Algorithms and Identifiers for the Internet X.509 Public Key Infrastructure Certificate and Certificate Revocation List (CRL) Profile,
    http://www.ietf.org/rfc/rfc3279.txt.
    RFC 3279 specifies OIDs and the structure for storing subject public key information for the RSA, DSA, DH and EC algorithms. The RFC also defines object identifiers and signature structures for hashing and signing algorithms. RFC 3279 is augmented by RFC 4055 and RFC 5480 to accommodate additional signature algorithms and schemes.

12. [RFC 3647]
    Public Key Infrastructure Certificate Policy and Certificate Practices Framework,
    http://www.ietf.org/rfc/rfc3647.txt.
    RFC 3647 presents a comprehensive list of topics that potentially need to be addressed in a certificate policy or a certification practice statement.

13. [RFC 4055]
    Additional Algorithms and Identifiers for RSA Cryptography for use in the Internet X.509 Public Key Infrastructure Certificate and Certificate Revocation List (CRL) Profile, http://www.ietf.org/rfc/rfc4055.txt.

RFC 4055 supplements RFC 3279, by describing the conventions for using the RSA-PSS signature algorithm, and the RSA-OAEP key transport algorithm.

14. [RFC 4251]
The Secure Shell (SSH) protocol Architecture, http://www.ietf.org/rfc/rfc4251.txt.
RFC 4251 specifies a protocol for secure remote login and other secure network services over an insecure network. This document describes the architecture of the SSH protocol, as well as the notation and terminology used in SSH protocol documents. It also discusses the SSH algorithm naming system that allows local extensions. The SSH protocol consists of three major components: The Transport Layer Protocol provides server authentication, confidentiality, and integrity with perfect forward secrecy; the User Authentication Protocol authenticates the client to the server; and the Connection Protocol multiplexes the encrypted tunnel into several logical channels. Details of these protocols are described in separate documents.

15. [RFC 6402]
Certificate Management over CMS (CMC), http://www.ietf.org/rfc/rfc5272.txt.
Formerly published as RFC 5272, RFC 6402 is a protocol standard for using certificate management services, such as enrollment, rekey, and revocation using PKCS #10 or the Cryptographic Message Syntax (CMS).

16. [RFC 5273]
Certificate Management over CMS (CMC): Transport Protocols,
http://www.ietf.org/rfc/rfc5273.txt.
RFC 5273 defines a number of transport mechanisms that are used to move CMC (Certificate Management over CMS (Cryptographic Message Syntax)) messages. The transport mechanisms described are: HTTP, file, mail, and TCP.

17. [RFC 5274]
Certificate Management Messages over CMS (CMC): Compliance Requirements,
http://www.ietf.org/rfc/rfc5274.txt.
RFC 5274 provides a set of compliance statements about the CMC (Certificate Management over CMS) enrollment protocol.

18. [RFC 5280]
Internet X.509 Public Key Infrastructure Certificate and Certificate Revocation List (CRL) Profile, http://www.ietf.org/rfc/rfc5280.txt.
RFC 5280 defines the formats for X.509 public key certificates and the corresponding CRLs. This RFC also defines the certificates, the certification paths, and the CRL processing rules.

19. [RFC 5295]
Specification for the Derivation of Root Keys from an Extended Master Session Key (EMSK), http://www.ietf.org/rfc/rfc5295.txt.

The Extensible Authentication Protocol (EAP) defines the Extended Master Session Key (EMSK) generation process. RFC 5295 defines how EMSK is used to derive root keys. Root keys are master keys that can be used for multiple security services, such as authentication, integrity, and encryption.

20. [RFC 5480]

Elliptic Curve Cryptography Subject Public Key Information,
http://www.ietf.org/rfc/rfc5480.txt.
RFC 5480 defines the format and structure of elliptic curve public keys in X509 certificates.

21. [RFC 5652]

Cryptographic Message Syntax (CMS),
http://www.ietf.org/rfc/rfc5652.txt.
RFC 5652 describes the Cryptographic Message Syntax (CMS) format. This syntax is used to digitally sign, digest, authenticate, or encrypt arbitrary message content.

22. [RFC 5914]

Trust Anchor Format, http://www.ietf.org/rfc/rfc5914.txt.
RFC 5914 describes a structure for representing trust anchor information. A trust anchor is an authoritative entity that is represented by a public key and associated data. The public key is used to verify digital signatures, and the associated data is used to constrain the types of information or actions for which the trust anchor is authoritative. The structure defined in this document is intended to satisfy format-related requirements defined in the Trust Anchor Management Requirements.

23. [RFC 5934]

Trust Anchor Management Protocol (TAMP), http://www.ietf.org/rfc/rfc5934.txt.
RFC 5934 defines a security protocol to securely update the trust anchor stores held by devices, equipment and applications.

23. [RFC 6024]

Trust Anchor Management Requirements, http://www.ietf.org/rfc/rfc6024.txt.
RFC 6024 describes some of the problems associated with the lack of a standard trust anchor management mechanism and defines requirements for data formats and push-based protocols designed to address these problems.

24. [SP 800-37]

SP 800-37-rev1: Guide for Applying the Risk Management Framework to Federal Information Systems – A Security Lifecycle Approach, February 2010
http://www.csrc.nist.gov/publications/PubsSPs.html.
SP 800-37-rev1 continues the evolution to a unified framework by transforming the traditional Certification and Accreditation process into the six-step Risk Management Framework (RMF). The revised process emphasizes: (i) building information security capabilities into Federal information systems through the application of state-of-the-

practice management, operational, and technical security controls; (ii) maintaining an awareness of the security state of information systems on an ongoing basis through enhanced monitoring processes; and (iii) providing essential information to senior leaders to facilitate credible decisions regarding the acceptance of risk to organizational operations and assets, individuals, other organizations, and the Nation arising from the operation and use of information systems.

25. [SP 800-38A]

   SP 800-38A: Recommendation for Block Cipher Modes of Operation - Methods and Techniques, December 2001, http://www.csrc.nist.gov/publications/PubsSPs.html.
   SP 800-38A defines five confidentiality modes of operation for use with an underlying symmetric key block cipher algorithm: Electronic Codebook (ECB), Cipher Block Chaining (CBC), Cipher Feedback (CFB), Output Feedback (OFB), and Counter (CTR). SP 800-38A is used with an approved block cipher algorithm. Compliance with this Recommendation is validated under the NIST Cryptographic Algorithm Validation Program (CAVP).

26. [SP 800-38B]

   SP 800-38B: Recommendation for Block Cipher Modes of Operation - the CMAC mode for Authentication, May 2005,
   http://www.csrc.nist.gov/publications/PubsSPs.html.
   SP 800-38B specifies a message authentication code (MAC) algorithm based on a symmetric key block cipher algorithm. This block cipher-based MAC algorithm, called CMAC, may be used to provide assurance of the authenticity and, hence, the integrity of binary data. Compliance with this Recommendation is validated under the NIST Cryptographic Algorithm Validation Program (CAVP).

27. [SP 800-38C]

   SP 800-38C Recommendation for Block Cipher Modes of Operation: the CCM Mode for Authentication and Confidentiality, May 2004,
   http://www.csrc.nist.gov/publications/PubsSPs.html.
   SP 800-38C defines a mode of operation, called CCM, for a symmetric key block cipher algorithm. CCM may be used to provide assurance of the confidentiality and the authenticity of computer data by combining the techniques of the Counter (CTR) mode and the Cipher Block Chaining-Message Authentication Code (CBC-MAC) algorithm[23]. Compliance with this Recommendation is validated under the NIST Cryptographic Algorithm Validation Program (CAVP).

28. [SP 800-38D]

   SP 800-38D: Recommendation for Block Cipher Modes of Operation: Galois/Counter Mode (GCM) and GMAC, November 2007,
   http://www.csrc.nist.gov/publications/PubsSPs.html.

---

[23] CBC-MAC is not currently an approved mode of operation except as a component of the CCM mode.

SP 800-38D specifies the Galois/Counter Mode (GCM) algorithm for authenticated encryption with associated data, and its specialization, GMAC, for generating a message authentication code (MAC) on data that is not encrypted. GCM and GMAC are modes of operation for an underlying, approved symmetric key block cipher algorithm. Compliance with this Recommendation is validated under the NIST Cryptographic Algorithm Validation Program (CAVP).

29. [SP 800-38E]

SP 800-38E: DRAFT Recommendation for Block Cipher Modes of Operation: The XTS-AES Mode for Confidentiality on Block-Oriented Storage Devices, January 2010, http://www.csrc.nist.gov/publications/PubsSPs.html.
SP 800-38E approves the XTS-AES mode of the AES algorithm by reference to IEEE Standard 1619-2007, subject to one additional requirement, as an option for protecting the confidentiality of data on block-oriented storage devices. The mode does not provide authentication of the data or its source.

30. [SP 800-38F]

SP 800-38F: Recommendation for Block Cipher Modes of Operation: Methods for Key Wrapping, December 2012, http://csrc.nist.gov/publications/PubsSPs.html.
SP 800-38F describes cryptographic methods that are approved for "key wrapping," i.e., the protection of the confidentiality and integrity of cryptographic keys.

31. [SP 800-53]

SP 800-53: Recommended Security Controls for Federal Information Systems, August, 2009, http://www.csrc.nist.gov/publications/PubsSPs.html.
SP 800-53 provides guidelines for selecting and specifying security controls for information systems supporting the executive agencies of the Federal government. The guidelines apply to all components of an information system that process, store, or transmit Federal information. The guidelines have been developed to help achieve more secure information systems within the Federal government by:

   a) Facilitating a more consistent, comparable, and repeatable approach for selecting and specifying security controls for information systems;

   b) Providing a recommendation for minimum security controls for information systems categorized in accordance with Federal Information Processing Standards (FIPS) 199, *Standards for Security Categorization of Federal Information and Information Systems*;

   c) Promoting a dynamic, extensible catalog of security controls for information systems to meet the demands of changing requirements and technologies; and

   d) Creating a foundation for the development of assessment methods and procedures for determining security control effectiveness.

32. [SP 800-53A]

SP 800-53A: Guide for Assessing Security Controls in Federal Information Systems – Building Effective Security Assessment Plans, June 2010, http://www.csrc.nist.gov/publications/PubsSPs.html.

SP 800-53A is intended to facilitate security control assessments conducted within an effective risk management framework. The assessment results provide organizational officials:

   a) Evidence about the effectiveness of security controls in organizational information systems;

   b) An indication of the quality of the risk management processes employed within the organization; and

   c) Information about the strengths and weaknesses of information systems that are supporting critical Federal missions and applications in a global environment of sophisticated threats.

33. [SP 800-56A]

SP 800-56A: Recommendation for Pair-Wise Key Establishment Schemes Using Discrete Logarithm Cryptography (Revised), May 2013, http://www.csrc.nist.gov/publications/PubsSPs.html.

SP 800-56A specifies key establishment schemes using discrete logarithm cryptography, based on standards developed by the Accredited Standards Committee (ASC) X9, Inc.: ANS X9.42 (Agreement of Symmetric Keys Using Discrete Logarithm Cryptography) and ANS X9.63 (Key Agreement and Key Transport Using Elliptic Curve Cryptography). Compliance with this Standard is validated under the NIST Cryptographic Algorithm Validation Program (CAVP).

34. [SP 800-56B]

SP 800-56B: Recommendation for Pair-Wise Key Establishment Using Integer Factorization Cryptography, August 2009, http://www.csrc.nist.gov/publications/PubsSPs.html.

SP 800-56B specifies key establishment schemes using integer factorization cryptography, based on ANS X9.44, Key Establishment using Integer Factorization Cryptography, which was developed by the Accredited Standards Committee (ASC) X9, Inc. Compliance with this Standard is validated under the NIST Cryptographic Algorithm Validation Program (CAVP).

35. [SP 800-56C]

SP 800-56C: Recommendation for Key Derivation through Extraction-then-Expansion, November 2011, http://www.csrc.nist.gov/publications/PubsSPs.html.

SP 800-56C specifies a two-step derivation procedure that employs an extraction-then-expansion technique for deriving keying material from a shared secret generated during a establishment scheme specified in [SP 800-56A] or [SP 800-56B]. Several application-specific derivation functions that use **approved** variants of this extraction-then-expansion procedure are described in NIST SP 800-135.

36. [SP 800-57-part1]
SP 800-57-part 1: Recommendation for Key Management – Part 1: General (Revised, July 2012,
http://www.csrc.nist.gov/publications/PubsSPs.html
SP 800-57 – Part 1 focuses on issues involving the management of cryptographic keys: their generation, use, and eventual destruction. Related topics, such as algorithm selection and appropriate key size, cryptographic policy, and cryptographic module selection, are also included.

37. [SP 800-57-part3]
SP 800-57-part 3: Recommendation for Key Management – Part 3: Application Specific Key Management Guidance, December 2009,
http://www.csrc.nist.gov/publications/PubsSPs.html
SP 800-57-part 3 is intended primarily to help system administrators and system installers adequately secure applications in common use. The guide also provides information for end users regarding application options left under their control in normal use of the application. Recommendations are given for a selected set of applications.

38. [SP 800-67]
SP 800-67: Recommendation for the Triple Data Encryption Algorithm (TDEA) Block Cipher, January 2012, http://www.csrc.nist.gov/publications/PubsSPs.html.
SP 800-67 specifies the Triple Data Encryption Algorithm (TDEA), including its primary component cryptographic engine, the Data Encryption Algorithm (DEA). Compliance with this Recommendation is validated under the NIST Cryptographic Algorithm Validation Program (CAVP).

39. [SP 800-89]
SP 800-89: Recommendation for Obtaining Assurances for Digital Signature Applications, November 2006,
http://www.csrc.nist.gov/publications/PubsSPs.html.
SP 800-89 specifies methods for obtaining the assurances necessary for valid digital signatures: assurance of domain parameter validity, assurance of public key validity, assurance that the key pair owner actually possesses the private key, and assurance of the identity of the key pair owner.

40. [SP 800-90A]
SP 800-90A: Recommendation for Random Number Generation Using Deterministic Random Bit Generators (Revised), January 2012,
http://www.csrc.nist.gov/publications/PubsSPs.html.
SP 800-90A specifies mechanisms for the generation of random bits using deterministic methods. The random bits may then be used directly or converted to

random numbers when required by applications using cryptography. The methods provided are based on hash functions, block cipher algorithms or number theoretic problems. Compliance with this Recommendation is validated under the NIST Cryptographic Algorithm Validation Program (CAVP).

80. [SP 800-102]

SP 800-102: Recommendation for Digital Signature Timeliness, September 2009, http://www.csrc.nist.gov/publications/PubsSPs.html.

SP 800-102 describes techniques for providing evidence of the time that a message was signed with a digital signature.

81. [SP 800-108]

SP 800-108: Recommendation for Key Derivation Using Pseudorandom Functions. October 2009, http://www.csrc.nist.gov/publications/PubsSPs.html.

SP 800-108 specifies several families of key derivation functions that use pseudorandom functions. These key derivation functions can be used to derive additional keys from a key that has been established through an automated key establishment scheme (e.g. as defined in [SP 800-56A] and [SP 800-56B]]), or from a pre-shared key (e.g., a manually distributed key).

82. [SP 800-115]

SP 800-115: Technical Guide to Information Security Testing and Assessment, September 2008, http://www.csrc.nist.gov/publications/PubsSPs.html.

SP 800-115 is a guide to the basic technical aspects of conducting information security assessments. The document presents technical testing and examination methods and techniques that an organization might use as part of an assessment, and offers insights to assessors on their execution and the potential impact they may have on systems and networks. For an assessment to be successful and have a positive impact on the security posture of a system (and ultimately upon the entire organization), elements beyond the execution of testing and examination must support the technical process. Suggestions for these activities—including a robust planning process, root cause analysis, and tailored reporting—are also presented in this guide.

83. [SP 800-126]

SP 800-126-r2: The Technical Specification for the Security Content Automation Protocol (SCAP): SCAP Version 1.2, September 2011, http://www.csrc.nist.gov/publications/PubsSPs.html.

SP 800-126-r2 defines the technical composition of SCAP Version 1.2 in terms of its component specifications, their interrelationships and interoperation, and the requirements for SCAP content. The technical specification for SCAP in this publication describes the requirements and conventions that are to be employed to ensure the consistent and accurate exchange of SCAP-conformant content and the ability to reliably use the content with SCAP-conformant products.

SCAP is a suite of specifications that standardize the format and nomenclature by which software flaw and security configuration information is communicated, both to machines and to humans. SCAP is a multi-purpose framework of specifications that support automated configuration, vulnerability and patch checking, technical control compliance activities, and security measurement. Goals for the development of SCAP include standardizing system security management, promoting interoperability of security products, and fostering the use of standard expressions of security content.

84. [SP 800-131A]

SP 800-131A: Transitions: Recommendation for Transitioning the Use of Cryptographic Algorithms and Key Lengths, January 2011, http://www.csrc.nist.gov/publications/PubsSPs.html.
SP 800-131A is intended to provide more detail about the transitions associated with the use of cryptography by Federal government agencies for the protection of sensitive, but unclassified information. The Recommendation addresses the use of algorithms and key lengths. (See also *Implementation Guidance for FIPS 140-2 and the CMVP Program,* G14 and G15, June 2012).

85. [SP 800-132]

SP 800-132: *Recommendation for Password-Based Key Derivation, Part 1: Storage Applications,* December 2010, http://www.csrc.nist.gov/publications/PubsSPs.html.
SP 800-132 specifies a family of password-based key derivation functions (PBKDFs) for deriving cryptographic keys from passwords or passphrases for the protection of electronically stored data or for the protection of data protection keys.

86. [SP 800-135]

SP 800-135 Revision 1: Recommendation for Existing Application-Specific Key Derivation Functions, December 2011, http://www.csrc.nist.gov/publications/PubsSPs.html.
SP 800-135 specifies security requirements for existing application-specific key derivation functions in several current security standards.

87. [SP 800-147]

SP 800-147: BIOS Protection Guidelines, April 2011, http://www.csrc.nist.gov/publications/PubsSPs.html.
SP 800-147 provides guidelines for preventing the unauthorized modification of *Basic Input/Output System (BIOS)* firmware on PC client systems. Unauthorized modification of BIOS firmware by malicious software constitutes a significant threat because of the BIOS's unique and privileged position within the PC architecture. A malicious BIOS modification could be part of a sophisticated, targeted attack on an

organization —either a permanent denial of service (if the BIOS is corrupted) or a persistent malware presence (if the BIOS is implanted with malware).

88. [TLS]
Various Transport Layer Security Related RFCs under
http://www.ietf.org/dyn/wg/charter/tls-charter.html.
These TLS RFCs describe how authentication, encryption, and integrity security services are provided for the HTTP packets. The RFCs cover the format of the security services payload for the packets, cipher suites for the security services, and key management techniques for the cryptographic algorithms used to provide the security services.

89. [X.509]
X.509: Information technology – Open Systems Interconnection – The Directory: Public-key and attribute certificate frameworks, IEC 9594-8.
This International Standard defines the formats for X.509 public key and attribute certificates and their associated CRLs, along with the certificate, certification path, and CRL processing rules.

90. [XML DSIG]
XML Signature Syntax and Processing (Second Edition), W3C Recommendation 10 June 2008, http://www.w3.org/TR/xmldsig-core.
XML DSIG describes the formats for digital signatures on XML documents, and for ancillary information (e.g., certificates, CRLs, Signer Identifiers, etc.) that can be used to assist in digital signature verification.

91. [XML ENC]
XML Encryption Syntax and Processing, W3C Recommendation 10 December 2002, http://www.w3.org/TR/xmlenc-core.
XML ENC describes the formats for encrypted XML documents and for ancillary information (e.g., certificates, CRLs, Recipient Identifiers, etc.) that can be used to assist in decryption.

Appendix B: Glossary of Terms

The following glossary contains the primary terms and definitions used in this Framework. Readers should also review the glossaries contained in [SP 800-57-part1].

Active State	The key lifecycle state in which a cryptographic key is available for use for a set of applications, algorithms, and security entities.
Algorithm Transition	The processes and procedures used to replace one cryptographic algorithm with another.
Anonymity	Assurance that public data cannot be associated with the owner in CKMS supported communications.
Application	A computer program designed and operated to achieve a set of goals or provide a set of services.
Archive (key and/or metadata)	To place an electronic cryptographic key and/or metadata into a long-term storage medium that will be maintained even if the storage technology changes. Also, the location where archived keys and/or metadata are stored.
Associated Metadata (also Metadata)	In the Framework, parameters used to describe properties associated with a cryptographic key that are explicitly recorded, managed, and protected by the CKMS.
Association Function	In this document, a function that protects a key and metadata from unauthorized modification and disclosure and authenticates the source of the metadata.
Audit	The procedures performed by an audit administrator to collect, analyze, and summarize the data required in a report to the system administrator regarding the security of the system.
Authoritative Time Source	A network entity that is relied upon to provide accurate time.
Backup (key and/or metadata)	To copy key and/or metadata to another facility so that the key and/or metadata can be recovered if the original values are lost or modified during operational usage.
CKMS	Policies, procedures, devices, and components designed to protect, manage, and distribute cryptographic keys and metadata. A CKMS performs cryptographic key management functions on behalf of one or more entities.
CKMS Component (Component)	Any hardware, software, or firmware that is used to implement a CKMS.

CKMS Device (Device)	Any combination of CKMS components that serve a specific purpose (e.g., firewalls, routers, transmission devices, cryptographic modules, and data storage devices).
CKMS Module	A logical entity that performs all required CKMS functions at a given location.
CKMS Profile	A document that provides an implementation independent specification of CKMS security requirements for use by a community of interest (e.g., U.S. Government; banking, health, and aerospace).
Commercial Off-The-Shelf (COTS)	Technology and/or a product that is ready-made and available for sale, lease, or license to the general public.
Compliant CKMS	A CKMS whose design specification addresses each requirement specified within this Framework.
Compromise	The unauthorized disclosure, modification, substitution or use of sensitive data (e.g., keys, metadata, and other security-related information).
Compromised State	A key lifecycle state in which a key is designated as compromised and is not to be used to apply cryptographic protection to data. Under certain circumstances, the key may be used to process already-protected data.
Cryptanalyze	To defeat cryptographic mechanisms, and more generally, information security services by the application of mathematical techniques.
Cryptographic Binding (Binding)	The use of one or more cryptographic techniques by a CKMS to establish a trusted association between a key and selected metadata elements.
Cryptographic Boundary	An explicitly-defined perimeter that establishes the boundary of all components of a cryptographic module.
Cryptographic Key (Key)	A string of bits, integers, or characters that constitute a parameter to a cryptographic algorithm.
Cryptographic Key Management System	A system for the management of cryptographic keys and their metadata (e.g., generation, distribution, storage, backup, archive, recovery, use, revocation, and destruction).
Cryptographic Module (Module)	A set of hardware, software and/or firmware that implements security functions (e.g. cryptographic algorithms and key establishment) and encompasses the cryptographic boundary.
Cryptographic Officer	An individual authorized to perform cryptographic initialization and management functions on the cryptographic components and devices of a CKMS.

Cryptography	The use of mathematical techniques to provide security services such as confidentiality, data integrity, entity authentication, and data origin authentication.
Cryptoperiod	The time span during which a specific key is authorized for use or in which the keys for a given system or application may remain in effect.
Deactivated State	The key lifecycle state in which a key is not to be used to apply cryptographic protection to data. Under certain circumstances, the key may be used to process already-protected data.
Designer	The person or organization having the ability, responsibility, and authority for specifying the devices comprising a new system and how the devices will be structured, coordinated, and operated.
Destroyed State	A key lifecycle state in which a key cannot be recovered or used.
Destroyed Compromised State	A key lifecycle state in which a key cannot be recovered nor used and is marked as compromised.
Security Domain (Domain)	A logical entity that contains a group of elements (e.g., people, organizations, information systems) that have common goals and requirements.
Entity	An individual (person), organization, device or process. An entity has an identifier to which it may be associated. (Sometimes called a party.)
Equivalent Security Domain Policies	Two domain security policies are equivalent if they permit the exchange of cryptographic keys from one security domain to another in a manner whereby the key is provided comparable protection in each domain.
Extensibility	A measure of the ease of increasing the capability of a system.
Firewall	The process integrated with a computer operating system that detects and prevents undesirable applications and remote users from accessing or performing operations on a secure computer.
Formal Language	A language whose syntax (i.e., rules for creating correct sentences with proper structure) is defined such that the rules are unambiguous and all syntactically correct sentences of the language can be recognized as being correct by an automaton (e.g., a computer running a syntax-analysis application program).
Framework	A description of the policies, procedures, components, and devices that are used to create a CKMS.
Garbled	The modification of data (e.g., a cryptographic key) in which one or more of its elements (e.g., bit, digit, character) has been changed or destroyed.

Generate Key	The key and metadata management function used to compute or create a cryptographic key.
Hardening	A process to eliminate a means of attack by patching vulnerabilities and turning off nonessential services. Hardening a computer involves several steps to form layers of protection.
Hash Value	The fixed-length bit string produced by a hash function.
Identifier	A text string used to indicate an entity (e.g., one that is performing a key management function) and by the CKMS access control system to select a specific key from a collection of keys.
Interoperability	A measure of the ability of one set of entities to physically connect to and logically communicate with another set of entities.
Key	See cryptographic key.
Key Agreement	A key establishment procedure where the resultant keying material is a function of information contributed by two or more participants, so that no entity can predetermine the resulting value of the keying material independently of any other entity's contribution.
Key Confirmation	A procedure to provide assurance to one entity (the key confirmation recipient) that another entity (the key confirmation provider) actually possesses the correct secret keying material and/or shared secret.
Key Distribution	The transport of a key and other keying material from an entity that either owns or generates the key to another entity that is intended to use the key. (Sometimes called key transport.)
Key Entry	The process by which a key (and perhaps its metadata) is entered into a cryptographic module in preparation for active use.
Key Establishment	The process by which a key is securely shared between two or more entities, either by transporting a key from one entity to another (key transport) or deriving a key from information shared by the entities (key agreement).
Key Label	A key label is a text string that provides a human-readable and perhaps machine-readable set of descriptors for the key. Hypothetical examples of key labels include: "Root CA Private Key 2009-29"; "Maintenance Secret Key 2005."
Key Lifecycle State	One of the set of finite states that describes the accepted use of a cryptographic key at a given point in its lifetime, including: Pre-Activation; Active; Suspended; Deactivated; Revoked; Compromised; Destroyed; Destroyed Compromised.

Key Output	The process by which a key (and perhaps its metadata) are extracted from a cryptographic module (usually for external storage).
Key Owner	An entity (e.g., person, group, organization, device, or module) authorized to use a cryptographic key or key pair.
Key Split	A parameter that, when properly combined with one or more other key splits, forms a cryptographic key.
Key State Transition	The process of moving from one key lifecycle state to another.
Key Transport	A key establishment procedure whereby one entity (the sender) selects and distributes the keying material to another entity (the receiver). Typically, key transport involves the use of cryptography to protect the keying material, but in some applications a trusted courier may be used instead. (Sometimes called key distribution.)
Key Update	The process used to replace a previously active key with a new key that is related to the old key.
Key Wrapping	A method of encrypting keys (along with associated integrity information) that provides both confidentiality and integrity protection using a symmetric key.
Keying Material	Key and/or metadata.
Least Privilege	The principle that each entity has access only to the information and resources necessary for legitimate use.
Malware	Software designed and operated by an adversary to violate the security of a computer (includes spyware, virus programs, root kits, and Trojan horses).
Metadata (also Associated Metadata)	In the Framework, parameters used to describe properties associated with a cryptographic key that are explicitly recorded, managed, and protected by the CKMS.
Metadata Element	One unit of metadata that is associated with a key and explicitly recorded and managed by the CKMS.
Mode of Operation	A set of rules for operating on data with a cryptographic algorithm and a key; often includes feeding all or part of the output of the algorithm back into the input of the next iteration of the algorithm, either with or without additional data being processed. Examples are: Cipher Feedback, Output Feedback, and Cipher Block Chaining.

Parameters	Specific variables and their values that are used with a cryptographic algorithm to compute outputs useful to achieve specific security goals.
Party	See entity
Pre-Activation State	A key lifecycle state in which a key has not yet been authorized for use.
Privacy	Assurance that the confidentiality of, and access to, certain information about an entity is protected.
Profile	A specification of the policies, procedures, components and devices that are used to create a CKMS that conforms to the standards of a customer sector (e.g., Federal, Private, or International).
Qubit	In quantum computing, a unit of quantum information – the quantum analogue of the classical bit.
Recover (General)	To get back; regain.
Recover (key and/or metadata)	To obtain or reconstruct a key and/or metadata from backup or archive storage.
Registration	The collection of procedures performed by a registration agent for verifying the identity and authorizations of an entity and establishing a trusted association of the entity's key(s) to the entity's identifier and possibly other metadata.
Rekey	The process used to replace a previously active key with a new key that was created completely independently of the old key.
Renewal	The process used to extend the validity period of a public key so that it can be used for an additional time period.
Revoked State	The key lifecycle state in which a previously active cryptographic key is no longer to be used to apply cryptographic protection to data.
Role	The set of acceptable functions, services, and tasks that a person or organization is authorized to perform within an environment or context.
Rootkit	Malware that enables unauthorized privileged access to a computer while actively hiding its presence from administrators by subverting standard operating-system functionality or other applications.

Router	A physical or logical entity that receives and transmits data packets or establishes logical connections among a diverse set of communicating entities (usually supports both hardwired and wireless communication devices simultaneously).
Scalability	The ability of a system to handle a growing amount of work in a capable manner or its ability to be enlarged to accommodate that growth.
Scheme	An unambiguous specification of a set of transformations that is capable of providing a (cryptographic) service when properly implemented and maintained. A scheme is a higher-level construct than a primitive and a lower level construct than a protocol.
Sector	A group of organizations (e.g., Federal agencies, private organizations, international consortia) that have common goals, standards, and requirements for a product, system, or service.
Security Domain	A collection of entities, including their CKMS, in which each CKMS operates under the same security policy — known as the Domain Security Policy.
Security Policy	The rules and requirements established by an organization that governs the acceptable use of its information and services, and the level and means for protecting the confidentiality, integrity, and availability of its information.
Security Strength	A number associated with the amount of work (that is, the base 2 logarithm of the minimum number of operations) that is required to cryptanalyze a cryptographic algorithm or system.
Semantics	The intended meaning of acceptable sentences of a language.
Standard	Something established by authority, custom, or general consent as a model or example.
Store (key and/or metadata)	To move a key and/or metadata into a medium from which the key and/or metadata may be recovered.
Suspended State	The key lifecycle state used to temporarily remove a previously active key from that status, but making provisions for later returning the key to active status, if appropriate.
Syntax	The rules for constructing acceptable sentences of a language.
Trust	A characteristic of an entity that indicates its ability to perform certain functions or services correctly, fairly, and impartially, along with assurance that the entity and its identifier are genuine.

Trust Anchor	One or more trusted public keys that exist at the base of a tree of trust or as the strongest link in a chain of trust and upon which a Public Key Infrastructure is constructed in a CKMS.
Trust Anchor Store	The location where trust anchor information is stored.
Trusted Association	The linking of a key with selected metadata elements so as to provide assurance that the key and its metadata are properly associated, originate from a particular source, have not been modified, and have been protected from unauthorized disclosure.
Unlinkability	Assurance that two or more related events in an information processing system cannot be associated with each other in CKMS-supported communications.
Unobservability	Assurance that an observer is unable to identify or make inferences about the parties involved in a transaction in CKMS-supported communications.
User	An individual authorized by an organization and its policies to use an information system, one or more of its applications, its security procedures and services, and a supporting CKMS.
Validate	To test cryptographic parameters or modules and confirm the test results to obtain assurance that the tested implementation is appropriate for use.
Validity Period	The lifespan of a public key certificate.

Appendix C: Acronyms

The following list contains acronyms used in the Framework.

ACS	Access Control System
AES	Advanced Encryption Standard
ANS	American National Standard
CBC	Cipher Block Chaining
CA	Certificate (Certification) Authority
CCM	Counter with Cipher Block Chaining-Message Authentication Code
CKL	Compromised Key List
CKMS	Cryptographic Key Management System(s)
CMS	Certificate Management System
COTS	Commercial Off-The-Shelf
CRL	Certificate Revocation List
DNSSEC	Domain Name System Security Extensions
EAP	Extensible Authentication Protocol
E-Mail	Electronic Mail
EC	Elliptic Curve
ECB	Electronic Codebook
EFS	Electronic File System
FIPS	Federal Information Processing Standard
FISMA	Federal Information Security Management Act
FR	Framework Requirement
fr	Framework Response
FT	Framework Topic
HMAC	Keyed-Hash Message Authentication Code
IDS	Intrusion Detection System
IKE	Internet Key Exchange
IP	Internet Protocol
IPSec	Internet Protocol Security
ISO/IEC	International Organization for Standardization/International Electrotechnical Commission
MAC	Message Authentication Code
NIST	National Institute of Standards and Technology
NTP	Network Time Protocol
OAEP	Optimal Asymmetric Encryption Padding
OCSP	Online Certificate Status Protocol
OFB	Output Feed Back
OID	Object Identifier

OMB	Office of Management and Budget
OTAR	Over-The-Air Rekeying
PKCS	Public Key Cryptographic Standards
PSS	Probabilistic Signature Scheme
RFC	Request For Comment
RSA	Rivest, Shamir and Adleman (Algorithm)
SCAP	Security Content Automation Protocol
S/MIME	Secure/Multipurpose Internet Mail Extensions
SP	Special Publication
SSH	Secure Shell
TDEA	Triple Data Encryption Algorithm
TLS	Transport Layer Security
VPN	Virtual Private Network
XML	Extensible Markup Language

www.ingramcontent.com/pod-product-compliance
Lightning Source LLC
Chambersburg PA
CBHW081500170526
45166CB00008B/2492